Electricity and Magnetism

Static Electricity, Current Electricity, and Magnets

Expanding Science Skills Series

BY

JOHN B. BEAVER, Ph.D., AND DON POWERS, Ph.D.

CONSULTANTS: SCHYRLET CAMERON AND CAROLYN CRAIG

COPYRIGHT © 2010 Mark Twain Media, Inc.

ISBN 978-1-58037-525-2

Printing No. CD-404122

Visit us at www.carsondellosa.com

Mark Twain Media, Inc., Publishers
Distributed by Carson-Dellosa Publishing Company, LLC

The purchase of this book entitles the buyer to reproduce the student pages for classroom use only. Other permissions may be obtained by writing Mark Twain Media, Inc., Publishers.

All rights reserved. Printed in the United States of America.   HPSO 218674

Table of Contents

Introduction 1	**Unit 6: Magnets**
How to Use This Book 2	Teacher Information 34
	Student Information 35
Unit 1: Historical Perspective	Quick Check 37
Teacher Information 3	Knowledge Builder 38
Student Information 4	Inquiry Investigation #1 39
Quick Check 6	Inquiry Investigation #2 40
Knowledge Builder 7	Inquiry Investigation #3 41
Knowledge Builder 8	Inquiry Investigation #4 42
	Inquiry Investigation #5 43
Unit 2: Electricity	
Teacher Information 9	**Unit 7: A Compass and Earth's Magnetic Field**
Student Information 10	Teacher Information 45
Quick Check 12	Student Information 46
Knowledge Builder 13	Quick Check 47
	Knowledge Builder 48
Unit 3: Current Electricity	
Teacher Information 14	**Unit 8: Electromagnetism**
Student Information 15	Teacher Information 49
Quick Check 16	Student Information 50
Knowledge Builder 17	Quick Check 51
Inquiry Investigation 18	Knowledge Builder 52
Unit 4: Static Electricity	**Unit 9: Electric Motors**
Teacher Information 20	Teacher Information 53
Student Information 21	Student Information 54
Quick Check 23	Quick Check 55
Knowledge Builder 24	Inquiry Investigation 56
Inquiry Investigation 25	
	Inquiry Investigation Rubric 60
Unit 5: Static Discharge	**National Standards** 61
Teacher Information 27	**Science Process Skills** 66
Student Information 28	**Definitions of Terms** 70
Quick Check 30	**Answer Keys** 73
Knowledge Builder 31	**Bibliography** 76
Inquiry Investigation 32	

Introduction

Electricity and Magnetism: Static Electricity, Current Electricity, and Magnets is one of the books in Mark Twain Media's new *Expanding Science Skills Series*. The easy-to-follow format of each book facilitates planning for the diverse learning styles and skill levels of middle-school students. The teacher information page provides a quick overview of the lesson to be taught. National science, mathematics, and technology standards, concepts, and science process skills are identified and listed, simplifying lesson preparation. Materials lists for Knowledge Builder activities are included where appropriate. Strategies presented in the lesson-planner section provide the teacher with alternative methods of instruction: reading exercises for concept development, hands-on activities to strengthen understanding of concepts, and investigations for inquiry learning. The challenging activities in the extended-learning section provide opportunities for students who excel to expand their learning.

Electricity and Magnetism: Static Electricity, Current Electricity, and Magnets is written for classroom teachers, parents, and students. This book can be used as a full unit of study or as individual lessons to supplement existing textbooks or curriculum programs. This book can be used as an enhancement to what is being done in the classroom or as a tutorial at home. The procedures and content background are clearly explained in the student information pages and include activities and investigations that can be completed individually or in a group setting. Materials used in the activities are commonly found at home or in the science classroom.

The *Expanding Science Skills Series* is designed to provide students in grades 5 through 8 and beyond with many opportunities to acquire knowledge, learn skills, explore scientific phenomena, and develop attitudes important to becoming scientifically literate. Other books in the series include *Chemistry, Simple Machines, Rocks and Minerals, Meteorology, Light and Sound,* and *Astronomy*.

The books in this series support the No Child Left Behind (NCLB) Act. The series promotes student knowledge and understanding of science and mathematics concepts through the use of good scientific techniques. The content, activities, and investigations are designed to strengthen scientific literacy skills and are correlated to the National Science Education Standards (NSES), the National Council for Teachers of Mathematics Standards (NCTM), and the Standards for Technological Literacy (STL). Correlations to state, national, and Canadian provincial standards are available at www.carsondellosa.com.

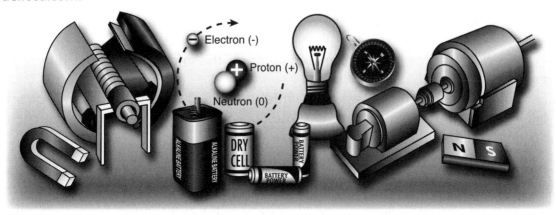

How to Use This Book

The format of *Electricity and Magnetism: Static Electricity, Current Electricity, and Magnets* is specifically designed to facilitate the planning and teaching of science. Our goal is to provide teachers with strategies and suggestions on how to successfully implement each lesson in the book. Units are divided into two parts: teacher information and student information.

Teacher Information Page

Each unit begins with a Teacher Information page. The purpose is to provide a snapshot of the unit. It is intended to guide the teacher through the development and implementation of the lessons in the unit of study. The Teacher Information page includes:

- National Standards: The unit is correlated with the National Science Education Standards (NSES), the National Council of Mathematics Standards (NCTM), and the Standards for Technological Literacy (STL). Pages 61–65 contain a complete list and description of the National Standards.
- Concepts/Naïve Concepts: The relevant science concepts and the commonly held student misconceptions are listed.
- Science Process Skills: The process skills associated with the unit are explained. Pages 66–69 contain a complete list and description of the Science Process Skills.
- Lesson Planner: The components of the lesson are described: directed reading, assessment, hands-on activities, materials lists of Knowledge Builder activities, and investigation.
- Extension: This activity provides opportunities for students who excel to expand their learning.
- Real World Application: The concept being taught is related to everyday life.

Student Pages

The Student Information pages follow the Teacher Information page. The built-in flexibility of this section accommodates a diversity of learning styles and skill levels. The format allows the teacher to begin the lesson with basic concepts and vocabulary presented in reading exercises and expand to progressively more difficult hands-on activities found on the Knowledge Builder and Inquiry Investigations pages. The Student Information pages include:

1. Student Information: introduces the concepts and essential vocabulary for the lesson in a directed reading exercise.
2. Quick Check: evaluates student comprehension of the information in the directed reading exercise.
3. Knowledge Builder: strengthens student understanding of concepts with hands-on activities.
4. Inquiry Investigation: explores concepts introduced in the directed reading exercise through labs, models, and exploration activities.

Safety Tip: Adult supervision is recommended for all activities, especially those where chemicals, heat sources, electricity, or sharp or breakable objects are used. Safety goggles, gloves, hot pads, and other safety equipment should be used where appropriate.

Unit 1: Historical Perspective
Teacher Information

Topic: Many individuals have contributed to the traditions of the science of electricity and magnetism.

Standards:
   NSES Unifying Concepts and Processes, (F), (G)
   See **National Standards** section (pages 61–65) for more information on each standard.

Concepts:
- Science and technology have advanced through contributions of many different people, in different cultures, at different times in history.
- Tracing the history of science can show how difficult it was for scientific innovations to break through the accepted ideas of their time to reach the conclusions we currently take for granted.

Naïve Concepts:
- All scientists wear lab coats.
- Scientists are totally absorbed in their research, oblivious to the world around them.
- Ideas and discoveries made by scientists from other cultures and civilizations before modern times are not relevant today.

Science Process Skills:
Students will be **collecting**, **recording**, and **interpreting information** while **developing the vocabulary to communicate** the results of their reading and research. Based on their findings, students will make an **inference** that many individuals have contributed to the traditions of science.

Lesson Planner:
1. Directed Reading: Introduce the concepts and essential vocabulary relating to the history of the science of electricity and magnetism using the directed reading exercise found on the Student Information pages.
2. Assessment: Evaluate student comprehension of the information in the directed reading exercise using the quiz located on the Quick Check page.
3. Concept Reinforcement: Strengthen student understanding of concepts with the activities found on the Knowledge Builder page. **Materials Needed:** Activity #1—colored poster board, scissors, tape, pencil; Activity #2—pyramid template, scissors, glue, pencil

Extension: Students research electrical safety and use the information to create a poster.

Real World Application: Understanding how electricity works is important to our safety. Water lowers the body's resistance to current. If your hands or feet are wet, never touch electrical devices, such as light switches, hair dryers, curling irons, mixers, or toasters.

Unit 1: Historical Perspective
Student Information

Our understanding of **static electricity**, **current electricity**, and **magnetism** is credited to many scientists. In examining their discoveries, we are able to see the connection and importance of these contributions to the science of electricity.

Static Electricity

Benjamin Franklin (1706–1790) conducted his famous kite experiment with lightning in June of 1752, over 257 years ago. He established a theoretical framework for the nature of electricity and electric charge. He observed static charges resulting from rubbing a hard rubber rod with rabbit fur and a glass rod with silk. He called the charges that were produced **negative** and **positive**, respectively. The practice of referring to a material that has gained electrons as having a negative charge and a material that has lost electrons as having a positive charge is still in use today.

Ben Franklin

William Gilbert

William Gilbert (1540–1603) observed that rubbing a piece of amber allowed it to attract lightweight materials. He is credited with giving the name *electricity* to this property of matter. The word electricity is derived from the Greek word for amber, *elektron*.

Dr. Luigi Galvani (1737–1798) of Bologna, Italy, worked with electrostatic devices in his laboratory and noted the influence that static electricity had on a frog's leg. The leg twitched when a scalpel he was using apparently conducted a charge from one of his machines to a nerve in the frog's leg. This observation led to the discovery of the electric battery.

Luigi Galvani

Allesandro Volta

Allesandro Volta (1745–1827) invented the battery in 1800 by layering silver and zinc plates separated by leather strips soaked in salt. This "pile" developed a fairly significant charge when the bottom silver and top zinc strips in the pile were connected or touched to metal pieces called **electrodes** that were placed in the water. A small amount of acid was added to the water to improve its conductivity. The process of using electricity to induce a chemical change was known as **electrolysis**. The use of electrolysis led to the discovery of several new elements.

Current Electricity

When electrons flow freely over a substance, we have what is called **electric current**. Some materials, such as most metals, conduct this flow of electrons more readily than other materials. Wires correctly connecting a flashlight battery and a lightbulb make a complete circuit. The glow of the lightbulb leads to the inference that electrons are flowing. The electrons are present in the wires, and the flashlight battery serves as the motive force to set them in motion. If you have a row of ball bearings in a plastic tube and roll an additional ball bearing into the tube, energy would be transferred from ball to ball along the length of the tube. This is analogous to how electrons flow through a wire. The electrons flow from one electrode in a battery, through the circuit, and into the other electrode on the battery. The flow of electrons is from the electrode on the battery labeled with the negative "–" sign and to the electrode labeled with the positive "+" sign.

Georg Ohm (1787–1854) developed the understanding of how voltage, resistance, and current are related and stated this relationship in Ohm's Law. Ohm started his career as an elementary school teacher, and later taught in secondary school. During his free time, he studied the factors that affected the flow of electricity across various metal conductors. **Ohm's Law** may be simply represented in the following mathematical equation: **E = IR**, where **E** stands for electromotive force or **volts**, **I** for electric current or **amperes**, and **R** stands for resistance or **ohms**. There is always some resistance to the flow of electricity, and Ohm's law provides a means for expressing the resistance, relative to voltage and amperage. The measurement of voltage is used to find the resistance in a system. A voltmeter is placed at any place in a circuit to find the drop in voltage.

Georg Ohm

Magnetism

Thales of Miletus (who lived in about 600 B.C.), a Greek scientist, noted the attractive force in a natural earth material called **lodestone** (leading stone), which is made up of iron ore. Observations of lodestone included noting that a slender piece of the material, when suspended from a string, oriented itself with the North and South Poles of the earth.

William Gilbert (1540–1603) published the first book dealing with magnetism in 1600. He noted that the earth acts as a giant magnet, and opposite poles attract each other.

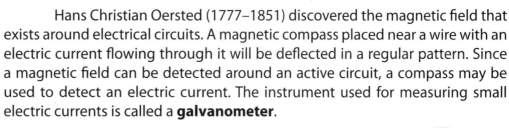
Thales of Miletus

Hans Christian Oersted (1777–1851) discovered the magnetic field that exists around electrical circuits. A magnetic compass placed near a wire with an electric current flowing through it will be deflected in a regular pattern. Since a magnetic field can be detected around an active circuit, a compass may be used to detect an electric current. The instrument used for measuring small electric currents is called a **galvanometer**.

Hans Christian Oersted

Michael Faraday (1791–1867) and Joseph Henry (1791–1878) discovered that a magnet moving in the vicinity of a coil of wire would generate an electric current. This is known as **induced current**. The large electric generators used today in power plants all over the world are based on the principles of induction outlined by Faraday and Henry.

Michael Faraday

Joseph Henry

Electricity and Magnetism Unit 1: Historical Perspective

Name: _____ Date: _____

Quick Check

Matching

_____ 1. Benjamin Franklin
_____ 2. Hans Christian Oersted
_____ 3. Thales of Miletus
_____ 4. William Gilbert
_____ 5. Allesandro Volta

a. published the first book dealing with magnetism
b. invented the battery in 1800
c. established a theoretical framework for the nature of electricity and electric charge
d. discovered the magnetic field that exists around electrical circuits
e. noted the attractive force in a natural earth material called lodestone

Fill in the Blanks

6. Dr. Galvani of Bologna, Italy, worked with electrostatic devices in his laboratory and noted the influence that _____ _____ had on a frog's leg.

7. The process of using electricity to induce a chemical change was known as _____.

8. When electrons flow freely over a substance, we have what is called _____.

9. The large electric generators used today in power plants all over the world are based on the principles of induction outlined by _____ _____ and _____ _____.

10. Georg Ohm (1787–1854) developed the understanding of how _____, _____, and _____ are related and stated this relationship in Ohm's Law.

Multiple Choice

11. The instrument used for measuring small electric currents is called a _____.
    a. lodestone
    b. galvanometer
    c. voltmeter
    d. magnet

12. A magnet moving in the vicinity of a coil of wire generates an electric current known as _____.
    a. induced current
    b. Ohm's Law
    c. static electricity
    d. magnetism

13. Lodestone is also called a _____.
    a. current
    b. galvanometer
    c. battery
    d. leading stone

Electricity and Magnetism Unit 1: Historical Perspective

Name: _____ Date: _____

Knowledge Builder

Activity #1: Scientist Book

Directions: Construct a book to display scientists and their contributions to the science of electricity and magnetism.

1. Cut two 12" circles out of poster board. Fold both circles in quarters, unfold them. Now, cut along one fold line to the center of the circle on both circles.

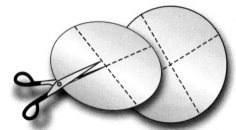

2. Starting at the slit, number one circle from 1 to 4 clockwise. Starting at the slit, number the second circle 5 to 8 clockwise.

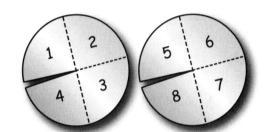

3. Place the numbered 5–8 circle on top of the numbered 1–4 circle and line up the slits.

4. Pull back the #8 section and tape the #5 and #4 sections together. Fold the quarter sections counterclockwise starting with the #8 section.

5. Fold until you have a pie slice. Using the information found on the Student Information pages, write and illustrate your book about scientists.

Electricity and Magnetism — Unit 1: Historical Perspective

Name: _____ Date: _____

Knowledge Builder

Activity #2: Pyramid

Directions: Cut out the pyramid template. Label sides: static electricity, current electricity, and magnetism. Draw an illustration that represents each type of electricity. List the names of the scientists that made a contribution to each area of electricity. On the base of the pyramid, write your name, and explain how electricity helps you in everyday life. Glue the tabs to the sides to complete your informational pyramid.

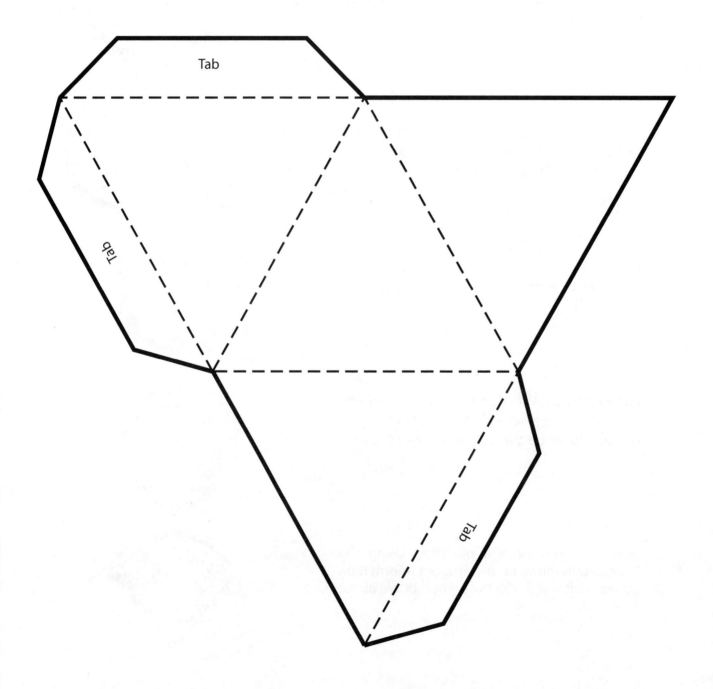

Electricity and Magnetism Unit 2: Electricity

Unit 2: Electricity
Teacher Information

Topic: Electricity is the interaction of electrical charges.

> **Standards:**
> **NSES** Unifying Concepts and Processes, (B), (F), (G)
> **NCTM** Numbers and Operations
> See **National Standards** section (pages 61–65) for more information on each standard.

Concepts:
- Electricity is the physical attraction and repulsion of electrons within and between materials.
- Electricity is measured in units of power called watts.

Naïve Concepts:
- Electricity flows like water in a pipe to the receiver.
- Electrons flow at the speed of light.
- Electricity vibrates in the wires until it is used.
- Electric companies supply electrons to your house.

Science Process Skills:
Students will be **collecting**, **recording**, and **interpreting information** while **developing the vocabulary to communicate** the results of their findings. Based on their findings, students will make an **inference** that there are many practical applications of electricity in the world around them.

Lesson Planner:
1. Directed Reading: Introduce the concepts and essential vocabulary relating to the science of electricity using the directed reading exercise found on the Student Information pages.

2. Assessment: Evaluate student comprehension of the information in the directed reading exercise using the quiz located on the Quick Check page.

3. Concept Reinforcement: Strengthen student understanding of concepts with the activities found on the Knowledge Builder page. **Materials Needed:** Activity #1—paper and pencil; Activity #2—home electricity meter, paper, pencil

Extension: Students research the number of watts used by the different appliances in their homes.

Real World Application: Many people pay an electricity bill each month. The amount paid depends on the number of kilowatt-hours of electricity used. The amount the electric company charges for each kilowatt-hour is called the rate.

Unit 2: Electricity
Student Information

Electricity is the flow of electrical charges. It is used to produce light and heat and to run motors. Some of the energy sources we use to make electricity are coal, natural gas, and oil. More environmentally friendly energy sources include sunlight, wind, and hydroelectric.

Electricity can be produced using turbines and generators. An energy source powers a turbine to run a generator. The **generator** turns large copper coils inside huge magnets, producing the electricity. A transformer sends the electric current to the power lines. The electricity is then carried to the users through wires.

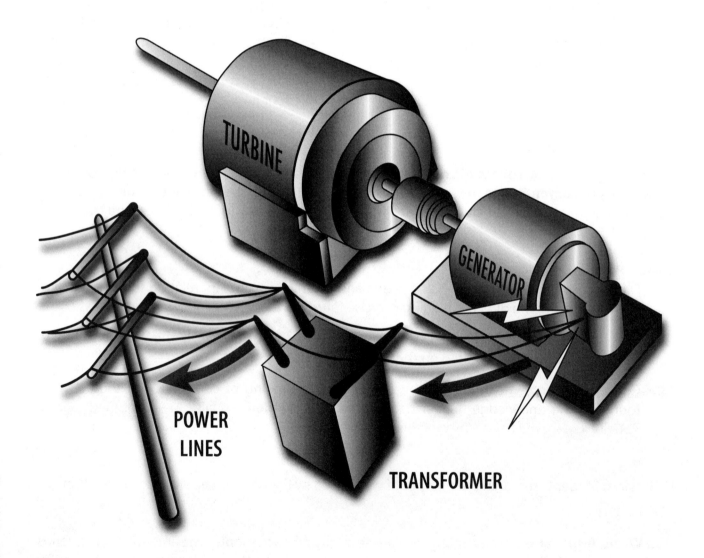

Electricity and Magnetism
Unit 2: Electricity

Electricity is the result of the movement of electrons. An electron is a small part of an atom. Every thing is made of matter. Matter is made up of tiny parts called **atoms**. Each atom has three even tinier parts. These parts are **protons**, **electrons**, and **neutrons**. The **nucleus** is the center of the atom. The protons and neutrons are small particles located in the nucleus of the atom. The protons have a positive (+) electrical charge. Neutrons are neutral. The neutrons have no charge. Protons and neutrons in an atom hold together, very tightly. Electrons are small particles in orbit around the nucleus. They are in orbit like the planets around the sun. This is the **Bohr Model** for atoms. The electrons have a negative (–) electrical charge.

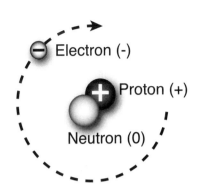

The protons and electrons of an atom are attracted to each other. They both carry an **electrical charge**: protons have a positive charge (+) and electrons have a negative charge (–). Like magnets, opposite charges attract each other. When an atom is in balance, it has an equal number of protons and electrons. When atoms are not balanced, they need to gain an electron. Electrons can be made to move from one atom to another. (A proton that has a positive charge attracts an electron that has a negative charge). When an electron moves between atoms, a current of electricity is created. As one electron is attached to an atom and another electron is lost, it creates a flow of electricity.

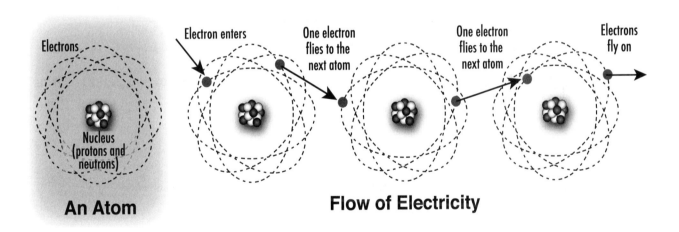

Electricity flows easily through some materials while other materials resist the flow of electricity. **Insulators** are materials such as plastic, rubber, cloth, and glass that hold electrons tightly. **Conductors** are materials that let electrons move more freely. Most metals are good conductors of electricity.

Electricity is measured in units called **watts**. One watt is a very small amount of power. A **kilowatt** represents 1,000 watts. A **kilowatt-hour** (kWh) is equal to the energy of 1,000 watts working for one hour. The amount of electricity a power plant generates or a customer uses over a period of time is measured in kilowatt-hours. The amount the electric companies charge for each kilowatt-hour is called the rate.

Electricity and Magnetism Unit 2: Electricity

Name: _____ Date: _____

Quick Check

Matching

_____ 1. electricity   a. small particles in orbit around the nucleus

_____ 2. protons   b. materials such as plastic, rubber, cloth, and glass

_____ 3. insulators   c. the flow of electrical charges

_____ 4. electrons   d. units used to measure electricity

_____ 5. watts   e. a positively charged particle located in the nucleus of an atom

Fill in the Blanks

6. The _____ turns large copper coils inside huge magnets, producing electricity.

7. The three parts of an atom are _____, _____, and _____.

8. The protons and electrons carry an _____ _____.

9. Metals are good _____ of electricity.

10. The amount of electricity a power plant generates or a customer uses over a period of time is measured in _____-_____.

Multiple Choice

11. An energy source powers a turbine to run a _____.

    a. motor   b. transformer

    c. generator   d. power plant

12. A good conductor of electricity is _____.

    a. glass   b. metal

    c. plastic   d. paper

13. When a(an) _____ moves between atoms, a current of electricity is created.

    a. proton   b. electron

    c. nucleus   d. neutron

Knowledge Builder

Activity #1: Cost of Electricity

Directions: A home owner's electricity bill depends on the number of kilowatt-hours used each month. Below is a list of common household appliances. For the following activity, the appliances listed below are shown with the watt-hours that might be used. (The number of watts used by a given household appliance will vary from different manufacturers.) Calculate how long it will take each appliance to use one kilowatt-hour.

Problem: A coffee maker uses 500 watts per hour. How long will it take for the coffee maker to use one kilowatt-hour? Divide 1,000 (1,000 watts equals one kilowatt-hour) by 500 (watts used per hour by the coffee maker).

Example: 1,000/500 = 2 hours

	Appliance	Watt-hours	Time
1.	Toaster	200 watts per hour	
2.	Clothes Dryer	2,000 watts per hour	
3.	Television	400 watts per hour	
4.	Refrigerator	250 watts per hour	

Activity #2: Meter Reading

Directions: The electric meter at your house measures how many kilowatt-hours (kWh) of electricity you use. Use the electric meter to work out how much electrical energy your home uses in a day. Take a reading of your electric meter. Record the reading along with the time below in the data table. The next day, at the same time, take another reading. Record the reading along with the time below in the data table. To find the kilowatt-hours used in one day, subtract the meter reading on Day #1 from Day #2, and record the difference in the data table. **(Caution: Electricity is dangerous! Please have an adult help you to take readings from your electric meter.)**

Day #1	Day #2	kWh Used in One day

Electricity and Magnetism Unit 3: Current Electricity

Unit 3: Current Electricity
Teacher Information

Topic: An electrical circuit is a complete path through which electrons flow.

Standards:
 NSES Unifying Concepts and Processes, (A), (B), (E)
 NCTM Geometry, Measurement, and Data Analysis and Probability
 STL Technology and Society; Abilities for a Technological World
 See **National Standards** section (pages 61–65) for more information on each standard.

Concepts:
- An electrical circuit is used to convert electrical energy into light, sound, and heat energy.
- Switches are used to open and close circuits.
- Two basic types of circuits are series and parallel.
- Electricity is the physical attraction and repulsion of electrons within and between materials.

Naïve Concepts:
- Current comes out from both poles of the battery and clashes in the bulb to light it.
- Charge flows through circuits at very high speeds. This explains why the lightbulb turns on immediately after the wall switch is flipped.
- Charge becomes used up as it flows through a circuit.

Science Process Skills:
Students will make **observations** and **inferences** about the flow of electricity in an electrical circuit. Students will create models of circuits and **explain** and **communicate** how the circuits operate.

Lesson Planner:
1. Directed Reading: Introduce the concepts and essential vocabulary relating to current electricity using the directed reading exercise found on the Student Information page.
2. Assessment: Evaluate student comprehension of the information in the directed reading exercise using the quiz located on the Quick Check page.
3. Concept Reinforcement: Strengthen student understanding of concepts with the activities found on the Knowledge Builder page. **Materials Needed:** Activity #1—flashlight blub, insulated wire, and a D-cell battery; Activity #2—a paper clip, 3" x 5" index card, two brass paper fasteners, flashlight blub, insulated wire, a D-cell battery.
4. Inquiry Investigation: Explore series and parallel circuits. Divide the class into teams. Instruct each team to complete the Inquiry Investigation pages.

Extension: Using a cardboard box, students create a "room" in their house. They wire a lightbulb and switch in the "room" so the light can be turned on or off.

Real World Application: Turning on the lights in a room or on a desk requires the use of a circuit. Radios, computers, and nearly all everyday electrical devices use circuits.

Unit 3: Current Electricity
Student Information

Current electricity is the movement of electrons, which creates the flow of electricity. An **electrical circuit** or closed circuit is a complete path through which electrons flow from an energy source, through a conducting wire and appliance, and back to the energy source. **Switches** are used to open and close circuits. An open circuit is off, and a closed circuit is on. Electrical circuits are used to convert electrical energy into light, sound, and heat energy. There are different ways to connect appliances in a circuit.

In its simplest form, a circuit must contain an energy source or battery and a piece of wire. Most often, the energy source is a dry-cell battery containing different chemicals that continuously react with each other, producing an excess of electrons. Hence, chemical energy is transformed into electrical energy. The chemicals commonly used in batteries are a mixture of ammonium chloride and zinc chloride, into which a carbon rod has been placed. The electrons leave the battery from the negative end of the battery (usually marked with a "–") and return to the positive end (usually marked with "+") through a wire called a conductor. Along the path of the wire, there may be one or more electrical appliances or devices, such as bulbs, motors, buzzers, etc.

Electrical circuits are either series or parallel. **Series circuits** require the electrical current to flow through all the devices in the circuit in just one path to make a complete circuit. **Parallel circuits** allow for more than one path along which the current can flow. Therefore, if one path in a parallel circuit has a break in it, the electrical current can still travel through the circuit that allows other devices in the circuit to operate. Because the current must flow through all the devices in a series circuit, the circuit has a higher resistance, and the bulbs in the circuit will be less bright. In a parallel circuit, because the current can flow around openings in the circuit, the resistance is less; therefore, the bulbs in the circuit are brighter.

Circuits are often shown with diagrams. To make diagrams consistent and easier to understand, electricians use symbols for certain objects. The illustration shows symbols that may be used to represent the electrical components in a circuit.

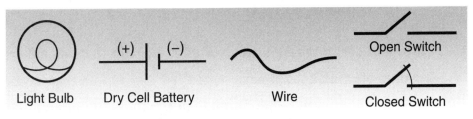

The flow of electricity in a circuit can be controlled by a number of different devices. A **switch** is used to open and close a circuit. It controls the flow of electricity in a circuit. As electricity moves through a circuit, heat is generated. If there is too much demand on a circuit due to too many devices, excess heat will be generated. This heat could melt the insulation on wires and cause nearby substances to start a fire. **Fuses** and **circuit breakers** are devices in a circuit that will either melt or trip to prevent heat build-up.

Electricity and Magnetism Unit 3: Current Electricity

Name: _____ Date: _____

Quick Check

Matching

_____ 1. current electricity
_____ 2. parallel circuit
_____ 3. series circuit
_____ 4. switches
_____ 5. electrical circuit

a. allows for more than one path along which the current can flow
b. electrical current flows through all devices in the circuit in one path
c. a closed circuit
d. used to open and close circuits
e. movement of electrons, which creates the flow of electricity

Fill in the Blanks

6. _____ _____ are used to convert electrical energy into light, sound, and heat energy.

7. _____ and _____ are devices in a circuit that will either melt or trip to prevent the build-up of heat.

8. Electrical circuits are either _____ or _____.

9. An electric circuit or closed circuit is a complete path through which _____ flow from an energy source, through a conducting wire and appliance, and back to the energy source.

10. A _____ must contain an energy source or battery and a piece of wire.

Multiple Choice

11. Fuses and circuit breakers are devices in a circuit that will either melt or trip to prevent the build-up of _____.
    a. ice
    b. dirt
    c. electrons
    d. heat

12. The electrons leave the battery from the _____ end of the battery (usually marked with a "–").
    a. negative
    b. positive
    c. closed
    d. open

Knowledge Builder

Activity #1: Simple Circuit

Directions: Electric current flows through a path called a circuit. Look at each diagram below. Using a flashlight bulb, insulated wire, and a D cell battery, try to light the bulb in the way shown in each picture. Circle whether the bulb will light or not for each setup.

1. yes/no   2. yes/no   3. yes/no   4. yes/no   5. yes/no   6. yes/no

Observation:

1. Which part of the bulb is the positive wire or the positive end of the battery touching? _____

2. One end of the wire touches the negative end (–) of the battery. Which part of the bulb is the other end of the wire touching? _____

Activity #2: The Switch

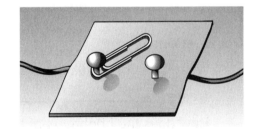

Directions: Make a hole in an index card for one paper fastener, approximately one-third of the way from the end of the index card. Insert a paper fastener through one end of the paper clip and attach the paper clip to the index card. Insert the second paper fastener into the index card in a location so it is touched by the other end of the paper clip. Now, the paper clip may be rotated to make contact with the second paper fastener. This will serve as a switch for your circuit. Attach a wire, approximately 10 cm in length, to each of the brass fasteners on the index card switch. Create a circuit with a battery and one bulb. Insert the switch between your battery and your bulb.

1. Close the switch in the circuit by moving the paper clip so that it touches both of the brass paper fasteners. What happens to the lightbulb when the switch is closed? _____

2. Open the switch in the circuit by moving the paper clip so it does not touch one of the brass paper fasteners. What happens to the lightbulb when the switch is open? _____

Electricity and Magnetism Unit 3: Current Electricity

Name: _____ Date: _____

Inquiry Investigation: Parallel and Series Circuits

Concepts:
- A parallel circuit is a circuit with two or more appliances that are connected so as to provide separate conduction paths for current for each appliance.
- A series circuit is a circuit with two or more appliances that are connected so as to provide a single conduction path for current.

Purpose: Create a circuit that will make a bulb brighter.

Procedure: Carry out the investigation. This includes gathering the materials, following the step-by-step directions, and recording the data.

Materials:
3 batteries (1.5 volts)   3 flashlight bulbs   bell wire   electrical tape

Experiment:
Part I
Step 1: Create a circuit with one bulb and 1 battery. Notice the brightness of the bulb.
Step 2: Using the materials provided, modify your circuit so the bulb is brighter.
Step 3: If the positive end (+) of the battery in the circuit touches the negative end (–) of a second battery, then your batteries are arranged in series. If the positive end of one battery is connected to the positive end of a second battery, and the negative ends of the batteries are connected in a similar way, then your batteries are connected in parallel. See the diagrams below to determine what type of circuit you have.

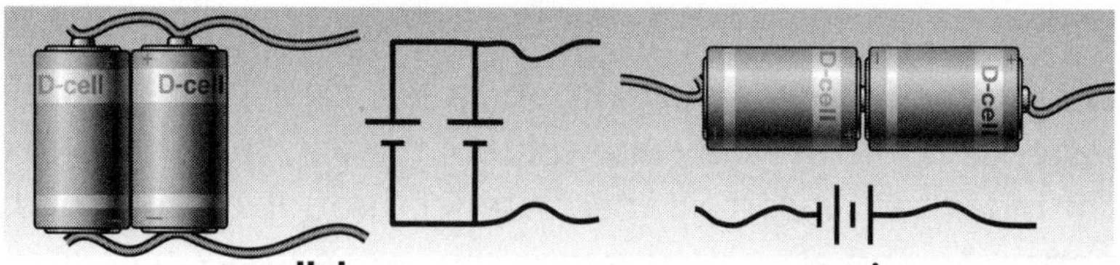

parallel   series

Observation:
1. What type of circuit have you created? _____
2. How does the brightness of the bulb compare to your original circuit with just one bulb and one battery? _____

Step 4: If the circuit you constructed in Part 1 was a series circuit, then construct a parallel circuit. If the circuit you constructed in Part 1 was a parallel circuit, then construct a series circuit.

Observation:
1. What type of circuit have you created now? _____
2. How does the brightness of the bulb compare to your original circuit with just one bulb and one battery? _____

Conclusion: What happens to the brightness of a bulb when two batteries are connected in a series vs. parallel circuit? _____

Electricity and Magnetism

Unit 3: Current Electricity

Name: _____ Date: _____

Part II

Step 1: Just like batteries, bulbs in a circuit may be arranged in series or parallel. Construct a circuit with two bulbs. (You may want to use two batteries in series in this circuit.) Once you have your circuit constructed and the bulbs working, disconnect one of the wires touching one of the bulbs.

Observation:

1. What happened? _____

2. Bulbs may be arranged in a series or parallel circuit. If two bulbs are in series, and one goes out, the other also goes out. If two bulbs are in parallel and one goes out, the other bulb remains lit. Are the bulbs in the circuit you have created in series or parallel? _____

Step 2: Below are diagrams of bulbs in a series and bulbs in a parallel circuit. If the parallel or series circuit you have created does not look like one of the circuits in the diagram below, then change your circuit now.

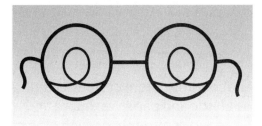
Diagram of bulbs in series

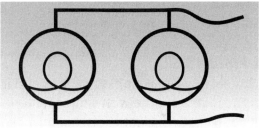
Diagram of bulbs in parallel

Step 3: If your circuit is series, disconnect one of the bulbs to see if the other bulb remains lit. If your circuit is parallel, disconnect one of the bulbs to see if the other bulb also goes out.

Step 4: Now investigate the other type of circuit.

Observation:

1. How does the brightness of the bulbs compare when in series and when in parallel?

_____

Conclusion:

1. Make a statement that explains what you have learned about bulbs in a series circuit.

_____

_____

2. Make a statement that explains what you have learned about bulbs in a parallel circuit.

_____

_____

Unit 4: Static Electricity
Teacher Information

Topic: Static charges involve the transfer of electrons from one material to another.

Standards:
   NSES Unifying Concepts and Processes, (A), (B), (G)
   NCTM Geometry, Measurement, and Data Analysis and Probability
   STL Technology and Society; Abilities for a Technological World
   See **National Standards** section (pages 61–65) for more information on each standard.

Concepts:
- Static electricity is electricity at rest.
- Static charges may be produced through friction.
- Static charges involve the transfer of electrons from one material to another.

Naïve Concepts:
- Materials that have lost an electron have lost them completely.
- All atoms are charged.
- Charged materials will only affect another charged object.

Science Process Skills:
Students will make **observations** and **inferences** about the nature of static charge and how materials acquire a charge with simple materials, such as plastic and paper, **estimate** and **measure** the relative strength of electric discharge; make **predictions** and **communicate** with others; **record**, **interpret**, and **analyze data**; **draw general conclusions**; and **make decisions**.

Lesson Planner:
1. Directed Reading: Introduce the concepts and essential vocabulary relating to static electricity using the directed reading exercise found on the Student Information pages.
2. Assessment: Evaluate student comprehension of the information in the directed reading exercise using the quiz located on the Quick Check page.
3. Concept Reinforcement: Strengthen student understanding of concepts with the activities found on the Knowledge Builder page. **Materials Needed:** Activity #1—balloon, sink with running water; Activity #2—balloon, watch, steamy bathroom
4. Inquiry Investigation: Explore static electricity. Divide the class into teams. Instruct each team to complete the Inquiry Investigation pages.

Extension: Students can experience static electricity by rubbing an inflated balloon on a wool sweater and then bringing the charged balloon near someone's hair.

Real World Application: Static electricity can build up when a person exits and re-enters a vehicle. A spark can discharge between your body and the fuel nozzle, igniting gasoline vapors around the fill spout.

Unit 4: Static Electricity
Student Information

Static electricity is electricity at rest. We are all familiar with the effects of walking across a carpet and touching a door knob or having a thin plastic bag stick to our clothing. These are examples of electric fields that are stationary or static, as opposed to a flowing charge or current electricity.

Electrostatic refers to electric charges that are confined to an object. This is also called **static electricity** or **electricity at rest**. It can also be defined as the charge-producing movement of electrons within a material and between materials. Electrons may move from atom to atom. Some materials hold electrons better than others. Plastic, cloth, and glass hold electrons tightly; they don't give them up. Other materials, such as metals, let electrons move more freely. It is important to remember that two atoms with the same charge will move apart, or **repel** each other, and two atoms with opposite charges will move together, or **attract** each other.

Understanding the basics of an atom, we can begin to see how static electricity works. All objects are made up of tiny parts, or **atoms**. Each atom has three tinier parts. These parts are **protons**, **electrons**, and **neutrons**. The protons and neutrons are located in the center of the atom. Electrons are in orbit around the nucleus. The protons have a positive (+) charge, and the electrons have a negative (-) charge. The neutrons are **neutral**, having no charge.

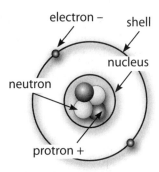

Static electricity is the result of the transfer of electrons. A **positive charge** forms when electrons are removed from a material, and a **negative charge** forms when electrons are added to a material. Objects can acquire a static charge through friction, conduction, and induction.

- **Friction:** Objects can acquire charges (electrons) through the rubbing of one surface against another. We experience this sometimes when we walk across a carpet in the winter in our socks.

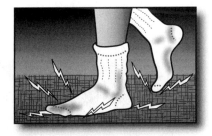

- **Conduction:** Charges (electrons) may be transferred through direct contact with other objects. This can happen when a person slides across a car seat and then touches a metal door handle.

- **Induction:** A charge (electrons) may be induced in other objects without contact, with no transfer of electrons. An induced charge is experienced when a charged object is brought near another material, and the material responds by movement. For instance, when you rub an inflated balloon on a wool sweater and then bring the charged balloon near someone's hair, the hair may respond by standing up straight. In this case, no electrons are transferred between the balloon and the hair; instead, free electrons are being repelled, and the resulting positive charge toward the ends of the hair causes it to be attracted to the negatively charged balloon.

It can be observed that objects with like charges repel each other, while objects with opposite charges attract each other, and a charged object attracts a neutral object. Other concepts that are related to static electricity charges include the following. Static electricity charges can be detected. The most common static electricity detector is the electroscope. An electrostatic field-meter can be used to measure the electrostatic field of an object in volts. Static electricity charges can be stored. When you shuffle your feet across a carpet, the friction causes a static charge to build up inside of you.

It is important to remember that static electricity demonstrations and experiments work best in dry air. You may have observed that you are more prone to static discharge experiences in the winter, when the relative humidity in indoor spaces is often very low. The reason for this is that charges naturally leak into the air, and humid air will cause the electrons to dissipate even more rapidly. Static electricity is more common in desert areas where the air is dry.

Static electricity is evident all around us, with lightning representing nature's most powerful display of static electricity. Lightning is a discharge of static electricity. Benjamin Franklin invented the lightning rod, which is used to protect wooden buildings from lightning damage. Lightning rods are attached to a building's highest point and connected to the ground by a thick wire, which transfers electrons from the ground to the sky. Sharp-pointed conductors, such as lightning rods, allow electrons to escape from a building's outer surfaces to the sky, instead of through the building.

Electricity and Magnetism Unit 4: Static Electricity

Name: _____ Date: _____

Quick Check

Matching

_____ 1. attract   a. move apart

_____ 2. humidity   b. rubbing

_____ 3. friction   c. electricity at rest

_____ 4. repel   d. move together

_____ 5. static electricity   e. water vapor in air

Fill in the Blanks

6. _____ have a positive (+) charge, and _____ a negative (–) charge.

7. When charges (electrons) are transferred through direct contact with other objects, it is called _____.

8. A _____ _____ forms when electrons are removed from a material, and a negative charge forms when electrons are added to a material.

9. Two atoms with opposite charges will move together, or _____ each other.

10. Benjamin Franklin invented the _____ _____, which is used to protect wooden buildings from lightning damage.

Multiple Choice

11. _____ is nature's most powerful display of static electricity.
    a. Current electricity   b. Lightning
    c. Conduction   d. Induction

12. The most common static electricity detector is the _____.
    a. lightning rod   b. carpet
    c. car seat   d. electroscope

13. Lightning rods allow _____ to escape from a building's outer surfaces to the sky, instead of through the building.
    a. neutrons   b. protons
    c. electrons   d. friction

14. Which of the following materials **does not** hold electrons tightly?
    a. plastic   b. cloth
    c. glass   d. metal

Electricity and Magnetism

Unit 4: Static Electricity

Name: _____ Date: _____

Knowledge Builder

Activity #1: Water Bending

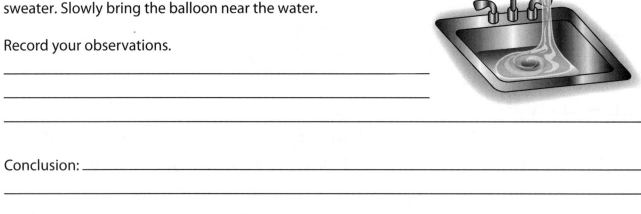

Directions: Turn the tap on so that you have a 1/8 inch thick stream of water flowing. Now briskly rub a balloon on a wool sweater. Slowly bring the balloon near the water.

Record your observations.
_____
_____
_____

Conclusion: _____
_____
_____
_____

Activity #2: Static Electricity and Bathrooms

Directions: Briskly rub a balloon on a wool sweater. Stick the balloon to a wall and time how long it stays before falling down. Record the results in the data table below. Next, briskly rub the balloon on a wool sweater. Now try sticking the balloon on a wall in a bathroom just after someone has taken a hot, steamy shower. Time how long it stays before falling down. Record the results in the data table below.

Type of Wall	Time (seconds)
Regular Wall	
Bathroom Wall	

Conclusion: _____
_____
_____
_____

Electricity and Magnetism Unit 4: Static Electricity

Name: _____ Date: _____

Inquiry Investigation: How an Object Can Become Statically Charged

Concept:
- Static charges may be produced through friction.
- Static charges can be transferred through direct contact between two objects, or indirectly by being brought near another object through induction.

Purpose: Explore how an object can become statically charged.

Procedure: Carry out the investigation. This includes gathering the materials, following the step-by-step directions, and recording the data.

Materials:
Styrofoam packing peanuts broken into smaller pieces or small shreds of paper
clear plastic containers such as plastic Petri dishes or plastic lids from food containers
a thin plastic bag, such as the bags found in the produce aisles of most grocery stores

Experiment:

Step 1: Place the shredded paper or packing peanuts under the clear plastic lid of a Petri dish or food container lid. It is best if the plastic lid is no more than 1–2 cm above the shredded material.

Step 2: Predict what you think will happen when you rub the plastic lid with the plastic bag. Record your prediction in the data table below.

Step 3: Now rub the lid with a piece of dry plastic bag. Record your observations in the data table below.

Results: Record your prediction and your observations below in the data table.

Predictions	Observations

Electricity and Magnetism　　　　　　　　　　　　　　　　　　　　　Unit 4: Static Electricity

Name: _____ Date: _____

Observation:

1. In this activity, there are three observable ways in which objects can become charged. What are you doing at the start of the activity that will produce a charge?
   _____
   _____
   _____

2. What are some other examples of static electricity that you have observed in your daily life?
   _____
   _____
   _____

3. What do you think caused the small bits of Styrofoam or paper to be attracted to the lid of the container?
   _____
   _____
   _____

4. How do you think the static environment affects the foam or paper bits? Is there a difference in the charge on the Styrofoam pieces on the lid as compared to those still on the table? Explain.
   _____
   _____
   _____

5. Why do some of the pieces of foam or paper fall off the lid?
   _____
   _____
   _____

Conclusion:
The three ways in which objects can become charged are through friction, induction, and conduction. Look at the diagram below and describe the three ways that a charge is produced in the activity.

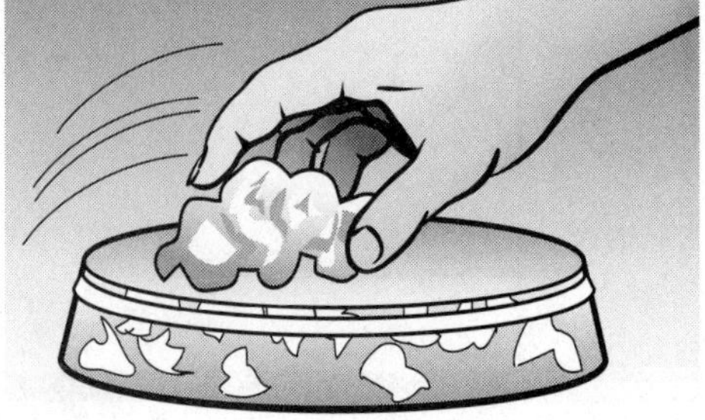

_____
_____
_____
_____

CD-404122 © Mark Twain Media, Inc., Publishers

Unit 5: Static Discharge
Teacher Information

Topic: Static charges involve the transfer of electrons from one material to another.

Standards:
NSES Unifying Concepts and Processes, (A), (B), (G)
NCTM Measurement and Data Analysis and Probability
STL Technology and Society; Abilities for a Technological World
See **National Standards** section (pages 61–65) for more information on each standard.

Concepts:
- Static charges may be produced through friction.
- Static charges are concentrated at the point of greatest curvature in a material, and if the material has sharp points, electrons may leak off the material into the air.

Naïve Concepts:
- Materials that have lost an electron have lost them completely.
- All atoms are charged.
- Charged materials will only affect another charged object.

Science Process Skills:
Students will make **observations** and **inferences** about the nature of static charge and discharge in an electrostatic generator called the electrophorus; **estimate** and **measure** the relative strength of electric discharge; make **predictions** and **communicate** with others; **record**, **interpret**, and **analyze data**; **draw general conclusions**; and **make decisions**.

Lesson Planner:
1. Directed Reading: Introduce the concepts and essential vocabulary relating to static discharge using the directed reading exercise found on the Student Information page.
2. Assessment: Evaluate student comprehension of the information in the directed reading exercise using the quiz located on the Quick Check page.
3. Concept Reinforcement: Strengthen student understanding of concepts with the activities found on the Knowledge Builder page. **Materials Needed:** Activity #1—aluminum pie pan, small piece of wool fabric, Styrofoam plate, pencil with a new eraser, thumbtack; Activity #2—cardboard, glass jar, scissors, clear tape, aluminum foil, comb
4. Inquiry Investigation: Explore static discharge. Divide the class into teams. Instruct each team to complete the Inquiry Investigation pages.

Extension: Students research lightning safety. Using the information they find, they create lightning safety posters.

Real World Application: Lightning is a giant discharge of electricity accompanied by a brilliant flash of light and a loud crack of thunder. Lightning is one of the leading weather-related causes of death and injury in the United States.

Unit 5: Static Discharge
Student Information

Electrostatic refers to electric charges that are confined to an object. This is also called "static electricity" or "electricity at rest." **Static discharge** is the release of **static** electricity when two objects come into contact. This is in contrast to **current electricity**, which is electricity that is moving through a circuit. Static charges are developed through friction between two or more objects or materials and may be generated through various means involving friction.

There are several types of **electrostatic generators**, including the Van de Graaff generator, the Wimshurst machine, and the electrophorus. All three devices produce charges as a result of friction and the migration of electrons within and between materials.

The **Wimshurst machine** uses a continuous-acting electrostatic device consisting of two glass disks with a large number of aluminum strips attached that rotate in opposite directions. As the disks are rotated, the metal strips act to induce and carry a charge. Pointed collector combs pick up and transmit the charges to two Leyden jars that store the charges temporarily.

The **Van de Graaff generator** consists of a large hollow metal sphere supported by an insulating cylinder. A wide belt made of insulating material runs from a pulley in the base of the generator over a second pulley at the generator's top. The drive pulley and belt are propelled by a motor, and the friction of the belt against a brush at the base of the generator causes a buildup of electrons on the belt. The electrons are picked up at the top of the generator by a second brush and transmitted to the metal sphere. The Van de Graaff generator is used in nuclear physics as a particle accelerator.

The **electrophorus** is the simplest electrostatic generator to set up. An **electrophorus** works by rubbing a Styrofoam sheet with a piece of wool. This causes the Styrofoam to become charged with electrons at the surface. Since Styrofoam is an **insulator** (material that does not conduct electricity), it cannot transfer its electrons to another material. A charge is built up at its surface, and when an aluminum pie pan is placed on the surface of the Styrofoam, a charge is induced in the pie pan. Note that the pie pan should have a Styrofoam cup taped to its center. The cup acts as an insulating handle, allowing one to pick up the pie pan. Since aluminum is a good **conductor** (material that carries electric current), the negative charge at the surface of the Styrofoam causes electrons in the pie

Electricity and Magnetism

Unit 5: Static Discharge

pan to migrate away from the Styrofoam toward the upper surface of the pie pan. If you bring your finger near the pie pan as it sits on the Styrofoam sheet, a visible charge will jump a narrow air gap between your finger and the pan as electrons are repelled toward the earth, or grounded. If the pan is then lifted off the foam sheet by its Styrofoam cup handle, and a hand is once again brought near the edge of the pan, a second discharge may be observed as the electrons return to the aluminum pan through the air from the finger. Theoretically, this may be repeated over and over again without much electrical charge loss from the Styrofoam base.

The pie pan does not build up an observable charge. The accumulation of charge in an object is greatest at the points of greatest curvature, so the largest discharges should be observed at the edges of a pie pan. A needle taped to the edge of the pan will further concentrate the charge and allow electrons to escape more readily.

Lightning is a dramatic natural example of static discharge. **Lightning** is an atmospheric discharge of electricity, which typically occurs during thunderstorms. Benjamin Franklin conducted his famous kite experiment with lightning in June of 1752, approximately 257 years ago. He established a theoretical framework for the nature of electricity and electric charge. For lightning to happen, there must be a spark. It is the same kind of spark that happens when sliding across a car seat. The spark occurs because of opposite charges.

In a storm cloud, small droplets of water make up the cloud. These small droplets of water are being carried up and down by air drafts in the cloud. The droplets of water rub against each other. The droplets with positive charges are lighter. They move to the top of the cloud. The droplets with negative charges are heavy. They move to the bottom of the cloud. When this happens, a spark may jump from the bottom of the cloud to the top. The spark creates a path for electricity. The path is from the bottom to the top of the cloud. Then we see a flash of lightning.

Using his knowledge of static discharge, Benjamin Franklin invented the **lightning rod**, a metal lightning conductor. Lightning rods are used to protect wooden buildings from lightning damage. They are attached to a building's highest point and connected to the ground by a thick wire, which transfers electrons from the ground to the sky. Sharp-pointed conductors, such as lightning rods, allow electrons to escape from a building's outer surfaces to the sky, instead of through the building.

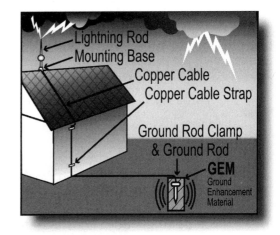

A glowing discharge of energy is often observed on the tips of ships' masts at night and on the trailing edges of airplane wings. Airplane wings are often equipped with pointed masts off the trailing edges of the wings to help remove the charge that the airplanes acquire in flight.

Electricity and Magnetism — Unit 5: Static Discharge

Name: _____ Date: _____

Quick Check

Matching

___ 1. conductor a. used to protect wooden buildings from lightning damage

___ 2. lightning rod b. material that carries electric current

___ 3. insulator c. electricity that is moving through a circuit

___ 4. electrostatic d. material that does not conduct electricity

___ 5. current electricity e. electric charges that are confined to an object

Fill in the Blanks

6. _____ _____ established a theoretical framework for the nature of electricity and electric charge.

7. The _____ generator consists of a large hollow metal sphere supported by an insulating cylinder.

8. The _____ machine uses a continuous-acting electrostatic device consisting of two glass disks with a large number of aluminum strips attached that rotate in opposite directions.

9. Sharp-pointed conductors, such as lightning rods, allow _____ to escape from a building's outer surfaces to the sky, instead of through the building.

10. Airplane wings are often equipped with pointed masts off the trailing edges of the wings to help remove the _____ that the airplanes acquire in flight.

Multiple Choice

11. Which of the following is NOT an electrostatic generator?
    a. Wimshurst   b. Van de Graaff
    c. Franklin   d. electrophorus

12. Aluminum is a good _____.
    a. conductor   b. insulator
    c. generator   d. electrophorus

13. The Van de Graaff generator is used in nuclear physics as a (an) _____.
    a. insulator   b. generator
    c. Leyden jar   d. particle accelerator

Electricity and Magnetism Unit 5: Static Discharge

Name: _____ Date: _____

Knowledge Builder

Activity #1: Lightning

Directions: Push a thumbtack through the center of an aluminum pie pan from the bottom. Push the eraser end of a pencil into the thumbtack. Place the Styrofoam plate upside-down on a table. Now, rub the plate with the wool for a couple of minutes. Pick up the aluminum pie pan using the pencil as a handle, and place it on top of the upside-down Styrofoam plate that you were just rubbing with the wool. Touch the aluminum pie pan with your finger. You should feel a slight shock. If you don't feel anything, try rubbing the Styrofoam plate again. Once you feel the shock, try turning the lights out before you touch the pan again. You should see a spark. (**Static electricity experiments work best on low-humidity days when the air is very dry.**)

Conclusion: Why does this activity produce a spark? _____
_____
_____
_____

Activity #2: Simple Electroscope

Directions: Turn a glass jar upside down onto a sheet of cardboard. Trace a circle around the jar. Draw the circle a little larger than the jar. Cut out the cardboard circle. Make two, 2-cm long slits 1 cm apart in the top of the cardboard circle. Cut an 11 cm x 2 cm strip of foil paper. Insert one end of the foil into one of the slits and the other end through the other slit. Carefully, insert all the foil into the slits so two even strips hang from the cardboard. Take a large piece of foil paper and cover the top of the cardboard circle. Place the cardboard circle on the mouth of the jar so that the strips of foil hang inside the jar. Rub a plastic comb through your hair and bring the comb in contact with the foil-covered cardboard lid. The foil strips that are inside the jar should move apart. (**Static electricity experiments work best on low-humidity days when the air is very dry.**)

Conclusion: Why do the foil strips inside the jar move apart? _____
_____
_____
_____

Electricity and Magnetism Unit 5: Static Discharge

Name: _____ Date: _____

Inquiry Investigation: An Electrophorus

Concepts:
- The electrophorus is a simple electrostatic generator that depends on an insulator and a conductor.
- Static charges are concentrated at the point of greatest curvature in a material, and if the materials have sharp points, electrons may leak off the material into the air.
- A needle may serve as a static electricity arrester when attached to a conducting material by allowing electrons to leak off the conduction material.

Purpose: Use an electrophorus to produce an electric charge.

Procedure: Carry out the investigation. This includes gathering the materials, following the step-by-step directions, and recording the data.

Materials:
30 centimeter square Styrofoam sheet   piece of wool cloth
masking tape   aluminum pie pan
Styrofoam cup   sewing needle

Experiment: (Works best on cool, dry days).

Step 1: Make loops of tape, adhesive side out, and secure the Styrofoam sheet to your desk top.
Step 2: Tape the Styrofoam cup to the inside of the aluminum pie pan to create an insulating handle for the pie pan.
Step 3: Rub the Styrofoam sheet rapidly with the piece of wool cloth. This causes electrons to be transferred from the wool cloth to the Styrofoam sheet. Be sure not to touch the Styrofoam sheet with your hand as you are rubbing it.
Step 4: Using the insulated handle, place the aluminum pie pan on the Styrofoam sheet.
Step 5: Predict what you think will happen if you bring your finger near the edge of the aluminum pie pan as it sits on the Styrofoam sheet. Record your prediction in the data table below.
Step 6: Now observe what happens when you touched the edge of the aluminum pan. Record your observations in the data table below.

Results: Record your prediction and your observations below in the data table.

Predictions	Observations

Electricity and Magnetism   Unit 5: Static Discharge

Name: _____ Date: _____

Step 7:   Follow the directions written in each box of the data table and record your observations.

Directions	Observations
A. After touching the pie pan and observing the result of this action, touch the pan a second and third time.	
B. Use the Styrofoam cup handle to lift the aluminum pan off the Styrofoam sheet. Touch the edge of the aluminium.	
C. After touching the pie pan and observing the result of this action, touch the pan a second and third time.	
D. Replace the aluminum pan on the Styrofoam sheet and touch it again.	
E. Repeat B and D several times. What do you observe?	

Step 8:   Use a piece of masking tape to attach a sewing needle to the edge of the aluminum pan, with the sharp end of the needle pointing out. Rub the Styrofoam sheet with the woolen cloth. Be sure not to touch the Styrofoam sheet with your hands as you are rubbing it.

Step 9:   Using the insulated handle, place the aluminum pie pan on the Styrofoam sheet. What happens when you touch the edge of the aluminum pan?
_____

Step 10: Now, use the Styrofoam cup handle to lift the aluminum pan off the Styrofoam sheet. Touch the edge of the aluminum pan. What did you observe?
_____

Conclusion:
1. Why is it important to handle the aluminum pan with the cup handle?
   _____

2. Explain what you think was happening when you touched the aluminum pan on the sheet and then touched it again when it was raised. _____
   _____

3. Explain what you think was happening when you touched the aluminum pan on the sheet then touched it again when it was raised with the needle attached.
   _____
   _____

Electricity and Magnetism Unit 6: Magnets

Unit 6: Magnets
Teacher Information

Topic: A magnetic field represents an area around a magnet that may influence other materials.

> **Standards:**
> **NSES** Unifying Concepts and Processes, (A), (B), (G)
> **NCTM** Measurement and Data Analysis and Probability
> **STL** Technology and Society; Abilities for a Technological World
> See **National Standards** section (pages 61–65) for more information on each standard.

Concepts:
- Materials may be classified as magnetic or nonmagnetic.
- Lines of force in a magnetic field can be visualized with a magnetic material, such as iron filings scattered over one or both poles of a magnet.

Naïve Concepts:
- All materials are attracted to a magnet.
- All silver-colored items are attracted to a magnet.
- All magnets are made of iron.

Science Process Skills:
Students will make **observations** and **inferences** about the nature of a magnetic field at each pole of a magnet, as well as the field that exists between unlike poles and two like poles; **estimate** and **measure** the relative strength of a magnetic field; make **predictions** and **communicate** with others; **record**, **interpret**, and **analyze data**; **draw general conclusions**; and **make decisions**.

Lesson Planner:
1. <u>Directed Reading</u>: Introduce the concepts and essential vocabulary related to magnets using the directed reading exercise on the Student Information pages.
2. <u>Assessment</u>: Evaluate student comprehension of the information in the directed reading exercise using the quiz located on the Quick Check page.
3. <u>Concept Reinforcement</u>: Strengthen student understanding of concepts with the activities found on the Knowledge Builder page. **Materials Needed:** Activity #1—magnets; Activity #2—plastic comb, a balloon, O-shaped dry cereal, tape, thread
4. <u>Inquiry Investigation</u>: Explore the influence of a magnetic field. Divide the class into teams. Instruct each team to complete the Inquiry Investigation pages.

Extension: Students grind up different dry breakfast cereals into very fine pieces. Then use a cow magnet or other strong magnet to attract iron from each of the cereals.

Real World Application: Magnetic attraction and magnetic fields have wide-ranging applications, including use in door closers, refrigerator door gaskets, motors, and generators.

Unit 6: Magnets
Student Information

A **magnet** is a device that attracts certain metals, such as iron, nickel, and cobalt. It can also attract or repel another magnet. The first magnets used by people were called lodestones. **Lodestone**, or iron ore, is also called magnetite and is found naturally on the earth's surface. It has distinctive magnetic qualities. People would carry a piece of lodestone (leading stone) on a string. The free-hanging magnet would point north. If they knew north, then it was easy to locate east, west, and south.

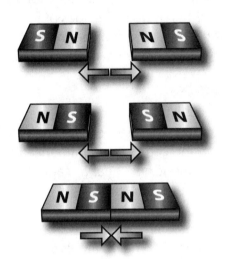

Magnets come in a variety of sizes, shapes, and strengths. One of the most common magnets is a bar magnet. Another is the horseshoe magnet. No matter what the shape is, the magnet will have two **poles**, or ends. One pole is a north pole; and the other is a south pole. **Magnetic force** is the attractive or repulsive force between the poles of magnets. If two magnets are placed near each other, the north pole of one will **attract** (move toward) the south pole of the other. If you place north poles toward each other, they will **repel** (move apart from) each other. If you place south poles toward each other, they will repel each other. Remember: like poles repel and unlike poles attract each other.

There are permanent magnets and temporary magnets. A **permanent magnet** is one that will hold its magnetic properties over a long period of time. Magnetite is a permanent magnet. Most permanent magnets we use are manufactured and are a combination or alloy of iron, nickel, and cobalt. A **temporary magnet** is one that will lose its magnetism.

A magnet can be used to make a magnet out of other metals. First, you must take the magnet and stroke the metal piece a few times in one direction. However, the metal piece is only a magnet for a short period of time, a temporary magnet. If a magnet is dropped, it may no longer be a magnet. Hit a magnet hard, and it may no longer be a magnet. Heating a magnet may also destroy it.

A magnet that is broken into two pieces will still have a north and a south pole. No matter how many times the magnet is broken; it will still have a north and a south pole. No one has been able to make a magnet with only one pole.

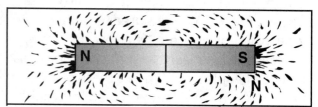

Anyone who has held two magnets in his/her hands and explored their interaction has felt the force that exists around magnets. This force through a distance is called the **magnetic field**. This invisible force exists around every magnet, and can be observed by using iron filings. When iron filings are scattered around a magnet, they always take the same shape. The curved lines of iron filings show the **magnetic field** or the space around a magnet where there is a magnetic force. The lines around the poles of the magnet are called **lines of force** or **lines of flux**. The lines of force indicate the direction of the magnetic field. The lines of force collectively are called the magnetic field. The stronger the magnet, the closer together the lines are at the poles. The force of a magnet is strongest near the poles.

A magnetic field can be observed in a number of ways. A magnetic compass can be used to estimate the magnetic field, a suspended magnet can be used as an indicator of the influence of the earth's magnetic field, and iron filings can be used to show the magnetic field near a magnet or between magnets.

How do magnets work? All objects are made of atoms. The atoms of some materials such as iron line up in a way that makes a magnet. The magnet is divided into a number of smaller regions called magnetic domains. A **magnetic domain** is a region in which the magnetic fields of atoms are grouped together and aligned. All the atoms in a domain point in the same direction. Each domain is essentially a tiny, self-contained magnet with a north and south pole with all the domains lined up to make one strong magnet. However, many materials like wood or plastic cannot be made into magnets. The atoms in these materials will not line up in order for the object to become a magnet.

unmagnetized *magnetized*

Electricity and magnetism are closely related. The movement of electrons causes both. Electricity comes from the flow of electrons from one atom to another. **Magnetism** is the alignment of electrons in the atom in the same direction making regions called domains, creating a magnetic field. Every electric current has its own magnetic field. This magnetic force can be used to make an electromagnet.

Electricity and Magnetism　　　　　　　　　　　　　　　　　　　　　Unit 6: Magnets

Name: _____ Date: _____

Quick Check

Matching

_____ 1. repel   a. force that surrounds a magnet
_____ 2. bar magnet   b. move apart from
_____ 3. magnetic field c. move toward
_____ 4. line of flux   d. common magnet
_____ 5. attract   e. magnetic force found at the poles of a magnet

Fill in the Blanks

6. No matter what the shape is, the magnet will have two _____, or ends.
7. No matter how many times the magnet is broken, it will still have a _____ and a _____ pole.
8. If two magnets are placed near each other, the _____ pole of one will attract the _____ pole of the other.
9. A _____ _____ is an attractive or repulsive force between the poles of magnets.
10. A _____ _____ is the space around a magnet where there is a magnetic force.

Multiple Choice

11. _____ is the alignment of electrons in the atom in the same direction making regions called domains, creating a magnetic field.
    a. Magnetism   b. Lodestone
    c. Compass   d. Domain

12. The force of a magnet is strongest near the _____.
    a. poles   b. sides
    c. bar magnet   d. horseshoe magnet

13. When iron filings are placed around the poles of a magnet, they _____ take the same shape.
    a. never   b. sometimes
    c. always   d. scatter

14. A _____ magnet is one that will lose its magnetism.
    a. domain   b. temporary
    c. permanent   d. field

Electricity and Magnetism

Unit 6: Magnets

Name: _____ Date: _____

Knowledge Builder

Activity #1: Magnetism

Directions: What can a magnet pick up? Find 10 objects in your classroom. Write them in the data table. Test the objects with a magnet. Record the results in the data table below.

Object	Magnetic?
1.	Yes/No
2.	Yes/No
3.	Yes/No
4.	Yes/No
5.	Yes/No
6.	Yes/No
7.	Yes/No
8.	Yes/No
9.	Yes/No
10.	Yes/No

Conclusion: What did you discover about magnetism? _____
_____
_____
_____

Activity #2: Electric Charge

Directions: Tie a 30-cm length of string to an O-shaped piece of dry cereal. Tape the other end to an edge of a table so the cereal can swing freely. Charge an inflated balloon by vigorously rubbing it on a wool cloth. Slowly, bring the balloon near the cereal. It will swing to touch the balloon. Hold the balloon still until the cereal jumps away. Now try to touch the balloon to the cereal again.

Observation: What happened when you brought the balloon near the cereal? _____
_____
_____

Conclusion: _____
_____
_____

Electricity and Magnetism Unit 6: Magnets

Name: _____ Date: _____

Inquiry Investigation #1: Observing the Influence of a Magnetic Field

Concept:
- Lines of force in a magnetic field can be visualized with a magnetic material.

Purpose: Observe the influence of a magnetic field.

Procedure: Carry out the investigation. This includes gathering the materials, following the step-by-step directions, and recording the data.

Materials:
sewing needle   plastic container (approximately 8 cm by 16 cm)
bar magnet   water
Styrofoam cube

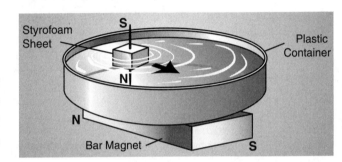

Experiment:
Step 1: Magnetize a sewing needle by stroking it along a bar magnet in one direction. Push the magnetized needle through the Styrofoam cube so it is perpendicular to the plastic container.
Step 2: Fill a plastic container (approximately 8 cm by 16 cm) with enough water so the needle floats vertically in the water.
Step 3: Set the plastic container with the floating needle over a bar magnet. Move the needle to the center of the container over the center of the bar magnet. Move the needle to different positions and observe. Rotate the bar magnet and try the needle in different positions. Record your observations.

Observation:
_____
_____

1. The direction of movement that is being observed in the needle represents a line of force (line of flux). Draw a diagram of the bar magnet and show the lines of force you observe.

2. Measure how far away from the bar magnet the force appears to extend. Record your measurement. _____

Electricity and Magnetism

Unit 6: Magnets

Name: _____ Date: _____

Inquiry Investigation #2: Observing a Magnetic Field

Concept:
- Lines of force in a magnetic field can be visualized with a magnetic material.

Purpose: Observe a magnetic field.

Procedure: Carry out the investigation. This includes gathering the materials, following the step-by-step directions, and recording the data.

Materials:
bar magnet   plastic sandwich bag   white paper   iron filings

Experiment:
Step 1:   Place a bar magnet in a plastic sandwich bag and lay it on a table.
Step 2:   Place a piece of paper over the magnet in the plastic bag.
Step 3:   Sprinkle iron filings over the paper on top of the plastic bag containing the magnet. Tap the paper. Record your observations by drawing a diagram of the observed pattern of iron filings.

Observation:

1. The observed pattern represents the magnetic field for the magnet. How far away from the magnet does the field extend? Measure and record this distance. _____

2. Where do most of the iron filings go? _____

3. How are your observations with the iron filings related to the floating needle in Inquiry Investigation #1? _____
_____

4. How does the observed pattern with the iron filings compare to the distance that the needle is attracted to the magnet? _____

5. How can a magnet attract the needle or the iron filings without touching them? _____
_____

6. What appears to be the strongest part of the magnet? _____

7. How could you use your iron filings to determine if other materials show a magnetic field?
_____

Conclusion: Magnetic field lines are used to study the direction of magnetic force and its relative strength. The magnetic field can also determine the extent of the force. Based on your observations, what conclusions can be made about the shape and extent of a magnetic field around a bar magnet?
_____
_____
_____

Electricity and Magnetism Unit 6: Magnets

Name: _____ Date: _____

Inquiry Investigation #3: Observe a Magnetic Force

Concept:
- Lines of force in a magnetic field can be visualized with a magnetic material.

Purpose: Observe magnetic attraction and force.

Procedure: Carry out the investigation. This includes gathering the materials, following the step-by-step directions, and recording the data.

Materials:
3 rubberized magnets   25 to 50 paper clips

Experiment:
Step 1: Form a hook with one of the paper clips and let the magnetic force hold the paper clip hook along the side of the magnet. (See diagram at the right.)
Step 2: Predict how many paper clips you think the magnet will hold. Record your prediction in the data table below.
Step 3: Carefully, place paper clips on the hook until the system fails. Record the number of paper clips the magnet held in the data table below.
Step 4: Repeat Step 3 two more times.
Step 5: Repeat Steps 1–4 with two magnets placed together, then three.
Step 6: Average the results for each trial and record your answers in the data table.

Results: Record the results for each trial in the data table below, then calculate the average for each trial. Then summarize your data on the lines below.

	1 Magnet	2 Magnets	3 Magnets
Prediction: # of paper clips magnet will hold			
Trial 1: # of paper clips magnet held			
Trial 2: # of paper clips magnet held			
Trial 3: # of paper clips magnet held			
Average: # of paper clips magnet held			

_____

_____

Conclusion: Magnetic force measurements could be used to compare the strength of different types of magnets. Based on your observations, what conclusions can be made about the force exerted by the different magnet combinations in this investigation? _____

_____

_____

Electricity and Magnetism Unit 6: Magnets

Name: _____ Date: _____

Inquiry Investigation #4: Measuring a Magnetic Force

Concept:
- Lines of force in a magnetic field can be visualized with a magnetic material.

Purpose: Observe the distance a magnet can move a paper clip.

Procedure: Carry out the investigation. This includes gathering the materials, following the step-by-step directions, and recording the data.

Materials:
plastic or wooden centimeter ruler   3 magnets
paper clips

Experiment:
Step 1:   Predict the distance at which a paper clip will be attracted to a magnet. Record your prediction in the data table below.
Step 2:   Place a magnet at one end of a plastic or wooden centimeter ruler and place a paper clip on the ruler. Slowly, slide the paper clip toward the magnet until the magnet attracts the paper clip. Record the distance in centimeters in the data table.
Step 3:   Repeat Step 2 two more times.
Step 4:   Repeat Steps 1–3 with two magnets stacked, then three.
Step 5:   Average the results for each trial and record the answers in the data table.

Results: Record the results for each trial in the data table below, then calculate the average for each trial. Then summarize your data on the lines below.

	1 Magnet	2 Magnets	3 Magnets
Prediction: distance paper clip will move (cm)			
Trial 1: actual distance paper clip moved (cm)			
Trial 2: actual distance paper clip moved (cm)			
Trial 3: actual distance paper clip moved (cm)			
Average: distance paper clip moved (cm)			

_____

_____

Conclusion: Magnetic force measurements could be used to compare the strength of different types of magnets. Based on your observations, what conclusions can be made about the force exerted by the different magnet combinations in this investigation? _____

_____

_____

Electricity and Magnetism Unit 6: Magnets

Name: _____ Date: _____

Inquiry Investigation #5: Strength of a Magnetic Force

Concept:
- Lines of force in a magnetic field can be visualized with a magnetic material.

Purpose: Observe the strength of magnetic force.

Procedure: Carry out the investigation. This includes gathering the materials, following the step-by-step directions, and recording the data.

Materials:
variety of objects in a classroom of a similar thickness: desk top, a classroom door, a glass window pane, a textbook, plastic storage box. (Do not test magnetic media such as CDs, DVDs, etc.)
magnets
paper clip

Part I

Experiment:

Step 1: Predict if a magnet or magnets will attract a paper clip through paper, wood, plastic, glass, and metal. Record your prediction in the data table below.

Step 2: Test each type of material to see if one or more magnets will attract a paper clip through the material. Record the information in the data table below.

Results: Record your predictions and test results below in the data table

	Prediction	Test Results
paper		
wood		
plastic		
metal		

Summary of your data for the materials tested:
_____
_____
_____
_____

Electricity and Magnetism Unit 6: Magnets

Name: _____ Date: _____

Part II

Experiment:

Step 1: After testing several different types of materials to determine if magnetic force will work through them, measure the thickness through which magnetic force is observable.

Step 2: Record the name of the material being tested in the data table below.

Step 3: Record the measured thickness in centimeters that magnetic force was observed in the boxes for one, two, and three magnets for each material in the data table below.

Results: Record the name of the material being tested and the measured thickness in centimeters that magnetic force was observed below in the data table.

	1 Magnet	2 Magnets	3 Magnets
Prediction: thickness through which a paper clip will be moved by magnetic force (cm)			
1st Material Tested:			
2nd Material Tested:			
3rd Material Tested:			
4th Material Tested:			

Summary of your data for the materials tested:

_____
_____
_____
_____
_____

Conclusion: Magnetic force measurements could be used to compare the strength of different types of magnets. Based on your observations, what conclusions can be made about the force exerted by the different magnet combinations in this investigation? _____

_____
_____
_____
_____

Electricity and Magnetism — Unit 7: A Compass and Earth's Magnetic Field

Unit 7: A Compass and Earth's Magnetic Field
Teacher Information

Topic: A compass needle points to magnetic north.

Standards:
NSES Unifying Concepts and Processes, (A), (D), (E)
NCTM Measurement and Data Analysis and Probability
STL Technology and Society; Abilities for a Technological World
See **National Standards** section (pages 61–65) for more information on each standard.

Concepts:
- Every magnet has two poles: north pole and south pole.
- A freely suspended magnet will align itself with the earth's magnetic poles.

Naïve Concepts:
- A compass needle points to the North Pole.

Science Process Skills:
Students will make **observations** and **inferences** about the nature of the earth's magnetic field and the interaction of a magnet with that field. A **model** of the earth showing its inferred magnetic field can be created using information from **observing** a magnetic field around a simple magnet.

Lesson Planner:
1. Directed Reading: Introduce the concepts and essential vocabulary relating to the nature of the earth's magnetic field using the directed reading exercise found on the Student Information page.
2. Assessment: Evaluate student comprehension of the information in the directed reading exercise using the quiz located on the Quick Check page.
3. Concept Reinforcement: Strengthen student understanding of concepts with the activities found on the Knowledge Builder page. **Materials Needed:** Activity #1—Styrofoam ball (baseball size or smaller), neodymium magnets, utility knife, glue, stapler with staples, colored permanent markers; Activity #2—bar magnet and plastic tub or glass pan (non-metallic container), Styrofoam or plastic plate (diameter must be longer than the length of the bar magnet and small enough to fit in a pan of water), magnetic compass

Extension: Teams of students create treasure hunts for other students using compass directions.

Real World Application: Magnetic compasses are used by surveyors, navigators, and hikers.

Unit 7: A Compass and Earth's Magnetic Field
Student Information

The earth acts as a giant magnet, complete with a North Pole and a South Pole. The circulation of Earth's liquid metal iron core creates a magnetic field. The north magnetic pole of the earth is not in the same location as the earth's geographic North Pole (axis of rotation). A suspended piece of magnetite will align itself with the magnetic poles of the earth.

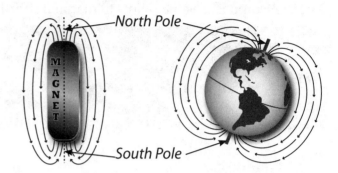

The North Pole on a globe is called **true north**. It is just under the North Star. If you followed a compass needle north, you would not end up at the North Pole. You would end up many miles from the North Pole. You would be at the magnetic north pole. When you looked up in the sky, there would be the North Star shining brightly hundreds of miles away. The North Star would be shining down on the North Pole, not the earth's magnetic north pole.

The magnetic north pole moves because of the circulation in Earth's liquid core. It will not always be where it is today. Today, the magnetic north pole is found near the Arctic Circle. It is about 500 miles from the true North Pole. It is slowly drifting across the Arctic. Scientists have plotted the movement. In the past few years, it has drifted about four degrees west. This is over 100 miles. Since 1900, it has moved 10 degrees north. It will keep moving. In fact, in a few years, it will be far from where it is today. At its present rate of movement, it could be in Russia within the next century.

A **compass** is a freely suspended magnet, usually in the form of a magnetized needle that will align itself with the earth's magnetic poles. Scientists refer to the end of a freely swinging magnet pointing to the earth's magnetic north pole as the "north-seeking pole," or simply the north pole of the magnet. Because the earth is a giant magnet, the compass needle will line up with Earth's magnetic field. The compass needle points to magnetic north, not the North Pole.

The compass has had a great impact on the human history of the earth. The invention of the compass and its use as a navigational device impacted trade, commerce, and the world's human migration patterns. Hans Christian Oersted (1777–1851) discovered the magnetic field that exists around electrical circuits. A magnetic compass placed near a wire with an electric current flowing through it will be deflected in a regular pattern. Since a magnetic field can be detected around an active circuit, a compass may be used to detect an electric current. A **galvanometer** is an instrument that uses a magnet to detect, measure, and determine the direction of small electric currents. Michael Faraday (1791–1867) and Joseph Henry (1791–1878) discovered that a magnet moving in the vicinity of a coil of wire would generate an electric current. This is known as induced current. The large electric generators used today in power plants all over the world are based on the principles of induction outlined by Faraday and Henry.

Electricity and Magnetism Unit 7: A Compass and Earth's Magnetic Field

Name: _____ Date: _____

Quick Check

Matching

_____ 1. true north

_____ 2. Hans Christian Oersted

_____ 3. compass

_____ 4. galvanometer

_____ 5. Michael Faraday and Joseph Henry

a. freely suspended magnet

b. detects, measures, and determines the direction of small electric currents

c. North Pole

d. discovered magnetic field that exists around electrical circuits

e. discovered that a magnet moving in the vicinity of a coil of wire would generate an electric current

Fill in the Blanks

6. The earth acts as a giant _____, complete with a North Pole and a South Pole.

7. The circulation of Earth's liquid metal iron core creates a _____ _____.

8. The north magnetic pole of the earth is not in the same location as the earth's _____ North Pole.

9. Scientists refer to the end of a freely swinging magnet pointing to the earth's magnetic north pole as the _____-_____ pole.

10. A magnetic compass placed near a wire with an electric current flowing through it will be _____ in a regular pattern.

Multiple Choice

11. The compass needle points to _____.
    a. the North Pole
    b. the South Pole
    c. magnetic north
    d. magnetic south

12. Today, the magnetic north pole can be found near _____.
    a. Russia
    b. the Arctic Circle
    c. Alaska
    d. the equator

13. The North Star shines down on the _____.
    a. magnetic north pole
    b. North Pole
    c. the Arctic Circle
    d. the equator

Knowledge Builder

Activity #1: Magnetic Earth

Directions: Using colored markers, draw the continents on a white Styrofoam ball creating a globe of the earth. Cut a slit into the side of the ball about half way through the ball. Make the slit to fit the size of a bar magnet. (You may need to use more than one magnet depending on the size of your Styrofoam ball.) Insert the magnet or magnets; with the north pole of the magnet pointing to the North Pole of your Styrofoam globe until you can attract a staple to the side of the ball. Glue the ball shut with the magnets inside. Remember that the magnet's north pole must correspond with the North Pole of your globe. Now, collect 20 to 30 bent staples; this can be done by stapling without paper. Place the staples around the equator and at the poles of your globe.

What do the staples represent? _____
_____
_____

Activity #2: Make a Compass

Directions: Fill a glass pan or plastic tub with enough water to float a Styrofoam plate. Place a bar magnet on the floating plate. Allow the plate to stop moving. It may be necessary to periodically move the plate away from the side of the pan or tub. After the plate has stopped moving, observe how the magnet has aligned itself. Use the magnetic compass to determine in which directions (north-south; east-west; northeast-southwest, etc.) the bar magnet points.

In which direction is the bar magnet pointing? _____

Rotate the plate with the bar magnet on it and allow it to turn freely again. Do the bar magnet and plate spin to point in the same direction as before? _____

Conclusion:
1. Explain why the bar magnet always pointed the same way. _____
_____
_____

2. Explain why the container holding the water should be made of either plastic or glass.
_____
_____

Unit 8: Electromagnetism
Teacher Information

Topic: An electromagnet takes advantage of the relationship between magnetism and electricity

Standards:
NSES Unifying Concepts and Processes, (A), (B), (E), (G)
NCTM Measurement
STL Technology and Society; Abilities for a Technological World
See **National Standards** section (pages 61–65) for more information on each standard.

Concepts:
- When a current of electricity is flowing through a coil of insulated wire wrapped around an iron core, it acts like a magnet.

Naïve Concepts:
- An electromagnet must have an iron nail.

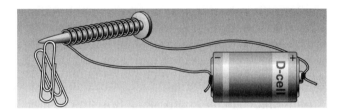

Science Process Skills:
Students will make **observations** and **inferences** about the creation of a magnetic field through induction in a coil of wire around an iron nail. Students will **measure** the relative strength of the electromagnet by making **observations** on how many paper clips the electromagnet is able to pick up. Students will also **compare** the relative strengths of several different electromagnets and **record** and **analyze the data** using data tables.

Lesson Planner:
1. Directed Reading: Introduce the concepts and essential vocabulary related to electromagnets using the directed reading exercise on the Student Information pages.
2. Assessment: Evaluate student comprehension of the information in the directed reading exercise using the quiz located on the Quick Check page.
3. Concept Reinforcement: Strengthen student understanding of concepts with the activities found on the Knowledge Builder page. **Materials Needed:** 2 D-cell batteries, 1 piece of insulated bell wire (1 meter long), 1 box of small paper clips, 2 battery holders, 1 large iron nail.

Extension: Students research electromagnets. Using the information they find, they create a list of common devices that use electromagnets.

Real World Application: Very large electromagnets are used in Germany and Japan to lift and drive trains called maglev trains. Maglev trains have no wheels. Maglev is short for magnetic levitation, which means that these trains float above special rails.

Unit 8: Electromagnetism
Student Information

An **electromagnet** is a temporary magnet in which the magnetic field is produced by the flow of electric current. Wrapping an insulated wire around a nail and then attaching the ends of the wire to a battery is a way to create a temporary magnet called an electromagnet. The use of magnets to induce electrical currents and the use of electric currents to produce magnetic fields are examples of the relationship between magnetic and electrical charges. The advantage of an electromagnet over a permanent magnet is that it can be turned on and off, and the strength of the magnetism can be varied.

Electricity and magnetism are closely related. The movement of electrons causes both. Electricity comes from the flow of electrons from one atom to another. **Magnetism** is the alignment of electrons in the atom in the same direction making regions called domains, creating a magnetic field. Every electric current has its own magnetic field. This magnetic force can be used to make an electromagnet.

Danish scientist Hans Christian Oersted (1777–1851) was the first person to prove that electricity and magnetism had something to do with each other. He discovered that electric currents create magnetic fields. In 1823, British scientist, William Sturgeon invented the electromagnet using a horseshoe-shaped piece of iron.

How do electromagnets work? All objects are made of atoms. The atoms of some materials such as iron can be made to line up in such a way as to make a magnet. The material of the core of the electromagnet is usually made of iron. The atoms inside of the iron core are divided into small areas called domains. The domains point in random directions. When a current is passed through the wire wrapped around the iron, the domains line up, creating a large magnetic field. When the current in the coil is turned off, the domains return to pointing in random directions. The magnetic field disappears.

Electromagnets have many applications, especially where magnetic contact is used to alternately turn an appliance on and off. The electric doorbell is an example of such a use. Pushing a doorbell switch completes an electric circuit that turns on an electromagnet that pulls the bell clapper toward a gong. This action breaks the circuit, turning off the electromagnet. The clapper moves back to a position that completes the circuit, starting the process over again. This cycle is repeated several times each second.

Electromagnets are used in recycling. Electromagnets are used on the end of crane booms for moving scrap metal. Additionally, electromagnetic sorting machines are used in separating iron-based material from other materials.

Electricity and Magnetism — Unit 8: Electromagnetism

Name: _____ Date: _____

Quick Check

Matching

_____ 1. Hans Christian Oersted

_____ 2. William Sturgeon

_____ 3. electromagnet

_____ 4. magnetism

_____ 5. electricity

a. proved that electricity and magnetism had something to do with each other

b. alignment of electrons in the atom in the same direction, making regions called domains

c. invented the electromagnet

d. temporary magnet in which the magnetic field is produced by the flow of electric current

e. comes from the flow of electrons from one atom to another

Fill in the Blanks

6. The advantage of an _____ over a permanent magnet is that it can be turned on and off, and the strength of the _____ can be varied.

7. _____ and _____ are closely related.

8. The material of the core of the electromagnet is usually made of _____.

9. Every electric current has its own _____ _____.

10. Wrapping an insulated wire around a nail and then attaching the ends of the wire to a battery is a way to create a temporary magnet called an _____.

Multiple Choice

11. Every electric current has its own _____.
    a. temporary magnet   b. magnetic field
    c. permanent magnet   d. electromagnet

12. Which of the following is NOT an example of electromagnet use?
    a. electric door bell
    b. electromagnetic sorting machines
    c. the end of crane booms for moving scrap metal
    d. turning lights on and off

13. The atoms inside of the iron core are divided into small areas called _____.
    a. electromagnets   b. magnetic fields
    c. domains   d. electric currents

Electricity and Magnetism   Unit 8: Electromagnetism

Name: _____   Date: _____

Knowledge Builder

Activity: Test the Strength of an Electromagnet

Directions: Using a piece of insulated bell wire one meter long and starting about 40 centimeters from one end, wrap the wire tightly around an iron nail 20 times. Connect the ends of the meter-long wire to two 1.5-volt batteries arranged in series. Touch the pointed end of the nail to a pile of paper clips. Now try using the nail head end to pick up the paper clips. In the data table below; record the number of paper clips you were able to pick up with the nail, comparing the pointed end of the nail and the nail head. Repeat the procedure above with 30 wraps, and then 40 wraps of wire around the nail.

	Number of Paper Clips Picked Up	
Twenty (20) Wraps	Pointed End	Nail head
Trial #1		
Trial #2		
Trial #3		

	Number of Paper Clips Picked Up	
Thirty (30) Wraps	Pointed End	Nail head
Trial #1		
Trial #2		
Trial #3		

	Number of Paper Clips Picked Up	
Forty (40) Wraps	Pointed End	Nail head
Trial #1		
Trial #2		
Trial #3		

Conclusion: What can you infer about the relationship of electricity to magnetism?
_____
_____
_____
_____
_____
_____

Unit 9: Electric Motors
Teacher Information

Topic: In a motor, electric energy is converted to mechanical energy.

> **Standards:**
> **NSES** Unifying Concepts and Processes, (A), (B), (E)
> **NCTM** Measurement
> **STL** The Nature of Technology; Technology and Society; Design; Abilities for a Technological World
> See **National Standards** section (pages 61–65) for more information on each standard.

Concepts:
- An electromagnet is the basis of an electric motor.
- In a motor, electric energy is converted to mechanical energy.

Naïve Concepts:
- Motors operate using gears and levers.

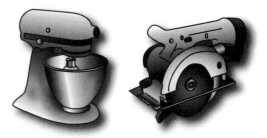

Science Process Skills:
Students will make **observations** about their motors as they **identify and control variables** that affect them; will **predict** how changing variables will affect their motors; will **make inferences** as to why the motor operates; will **measure** changes in their motors as variables change; and will **experiment** by **asking questions, manipulating materials,** and **analyzing the results** of their experiments.

Lesson Planner:
1. <u>Directed Reading</u>: Introduce the concepts and essential vocabulary relating to electric motors using the directed reading exercise found on the Student Information page.
2. <u>Assessment</u>: Evaluate student comprehension of the information in the directed reading exercise using the quiz located on the Quick Check page.
3. <u>Inquiry Investigation</u>: Explore electric motors. Divide the class into teams. Instruct each team to complete the Inquiry Investigation pages.

Extension: Students walk through their house and list all the electric motors.

Real World Application: Electric cars use an electric motor that runs on batteries that can be recharged. Generally, the more batteries there are, the farther you can drive the car on one charge.

Unit 9: Electric Motors
Student Information

An **electric motor** is a device that converts electricity to mechanical work. Electric motors are everywhere! Almost every mechanical movement that you see around you is caused by an electric motor. By understanding how a motor works, you can learn a lot about magnets, electromagnets, and electricity in general.

In an electric motor, electricity passes through an armature, producing magnetic fields nearby. An **armature** is the rotating part of an electric motor that mostly consists of coils of wire around a metal core. The armature is surrounded by **permanent magnets**, or magnets that stay magnetized. Once the coil begins to spin, the two magnetic fields interact with each other. During one half of the **rotation**, or turn of the coil, the magnetic forces of the electric coil and the permanent magnet are repelling each other. When the force of attraction and repulsion are equal, the coil will spin. Additional gears, pulleys, or other mechanical devices may be added to the ends of the spinning armature to allow the motor to drive a piece of equipment.

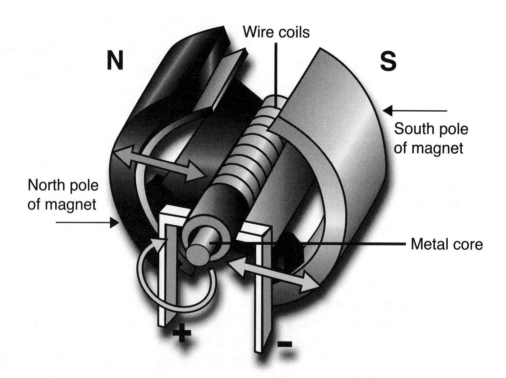

Electric motors are used in many ways in our daily lives. Electric motors in cars run everything from the ventilation fan to the power windows and seats and from the windshield wipers to the alternators and starters under the hood. Small electric motors are also used to operate the turning of the CDs in your CD players and computers. Large electric motors are used in electrical generating stations to produce electricity for everyday use. Research is ongoing to develop new and more efficient motors for use in **vehicles**, or means of transportation, and appliances.

Electricity and Magnetism Unit 9: Electric Motors

Name: _____ Date: _____

Quick Check

Matching

_____ 1. rotation   a. magnets that stay magnetized

_____ 2. vehicles   b. turn

_____ 3. armature   c. device that converts electricity to mechanical work

_____ 4. permanent magnets   d. means of transportation

_____ 5. electric motor   e. rotating part of an electric motor

Fill in the Blanks

6. Additional _____, _____, or other mechanical devices may be added to the ends of the spinning armature to allow the motor to drive a piece of equipment.

7. The armature is surrounded by _____ magnets.

8. When the force of _____ and _____ are equal, the coil of an armature will spin.

9. Large electric motors are used in electrical generating stations to produce _____ for everyday use.

10. Research is ongoing to develop new and more _____ motors for use in vehicles and appliances.

Multiple Choice

11. An armature is the rotating part of an electric motor that mostly consists of coils of _____ around a metal core.
    a. magnets
    b. wire
    c. electromagnets
    d. aluminum

12. The armature is surrounded by _____.
    a. temporary magnets
    b. electricity
    c. permanent magnets
    d. gears

13. A(An) _____ is a device that converts electricity to mechanical work.
    a. magnet
    b. pulley
    c. electromagnet
    d. electric motor

Electricity and Magnetism Unit 9: Electric Motors

Name: _____ Date: _____

Inquiry Investigation: Build an Electric Motor

Concept:
- By supplying electric current to a conductor in a magnetic field, electric energy is converted into mechanical energy.

Purpose: Build a simple working electric motor.

Procedure: Carry out the investigation. This includes gathering the materials, following the step-by-step directions, and recording the data.

Materials:

1 meter of #28 enameled copper wire
2 ring magnets
sandpaper
bell wire for circuit connections
clay

2 large paper clips
2 D-cell flashlight batteries
masking tape
9–10 oz. plastic cup

Experiment:

Making the Armature

Step 1: To construct the armature, or coil, of the motor, begin with approximately one meter of #28 enameled copper wire. Leaving an end of about 10 centimeters of wire, begin to wrap the remaining piece of wire tightly around the circular part of a D-sized battery. Wrap the wire seven times around the battery.

Step 2: Slip the loops of wire off the battery and wrap one end of the wire 3-4 times around the coil of wire to hold all seven loops together and to keep its circular shape.

Step 3: Grip the other end of the wire and wrap it tightly around the coil opposite the first location in the same manner.

Step 4: Bend each end of the wire so it sticks straight out from each coil opposite the other. It is important that the loop is symmetrical with the ends of wire opposite each other. See the diagram below (magnified view of looped wire).

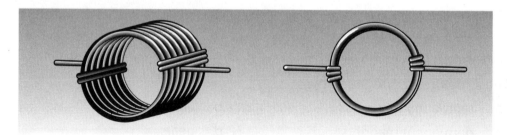

Step 5: Trim the ends of the wire sticking straight out to a length of about 5 centimeters each.

Step 6: Use the sandpaper to scrape the enamel off the wire from the coil to the wire's end. This can be easily done by gripping the wire ends of the coil with the sandpaper and pulling the wire through your fingers while holding the wire with the sandpaper. Repeat this action numerous times. (Close examination of the wire will reveal that the wire is shiny, and the enamel is removed.)

Electricity and Magnetism

Unit 9: Electric Motors

Name: _____ Date: _____

Making the Armature Holders

You must now construct the holders for your armature or coil. Two possible ways to do this are as follows:

The Cup Holder

Step 1: Using two jumbo-sized paper clips, open the paper clips into an "S" shape. Now, take the smaller end of the paper clip and twist it 180 degrees, forming a small hole where the paper clip would normally bend. Do the same with the other paper clip.

Step 2: Turn a plastic cup upside-down and tape one paper clip on one side of the cup with the hole towards the top, and the middle of the paper clip level with the edge of the inverted cup. Repeat the procedure with the other paper clip on the opposite side of the cup so that the two holes you formed by bending the paper clips are level. It is important that the large paper clips be fairly symmetrical with each other. (See diagram A.)

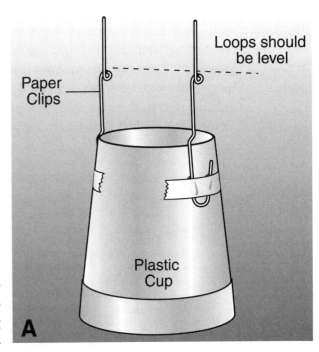

The Table-Top Holder

Step 1: Using two large-size paper clips, open one to 90 degrees.

Step 2: Take the larger end of the paper clip and turn it 180 degrees in its natural direction to form a small loop.

Step 3: Do the same with the other paper clip.

Step 4: Tape the two 90-degree paper clips to the table top, approximately 8 centimeters apart. The paper clips may be taped to a rigid piece of cardboard for portability. It is important that the large paper clips be fairly symmetrical with each other (See diagram B).

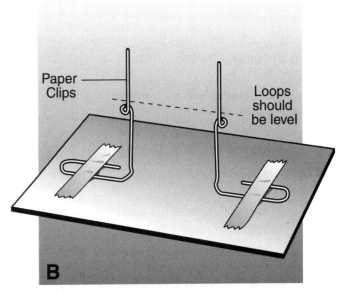

Electricity and Magnetism Unit 9: Electric Motors

Name: _____ Date: _____

Connect Armature to Armature Holder

Step 1: Lay the armature you constructed flat on the table. It is important that the armature is as balanced as possible. Be sure the two wire ends are flat on the table, and are opposite each other, and point in opposite directions.

Step 2: Gently, slip one end of the wire into the loop you made on one of the paper clips. Then slip the other wire into the other paper clip's loop.

Step 3: Gently, spin the wire coil to check its balance; it should spin without any wobble. If there is wobbling, then gently bend the ends of the coil to make it balance.

Step 4: Place a small permanent magnet on the cup or table directly below the coil.

Step 5: Using bell wire, connect a fresh 1.5-volt battery, positive end to one paper clip and negative end to the other. The armature may begin to spin or wobble. If it begins to spin, CONGRATULATIONS, you have created a simple electric motor!

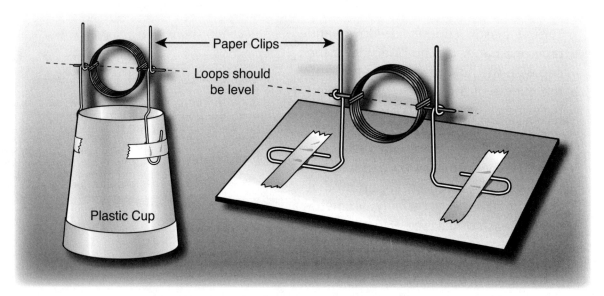

Troubleshooting

This process may be somewhat frustrating at first, but continue to work with the motor. Getting your armature to spin is a rewarding experience.

1. If the armature moves but does not spin completely around, gently tap the armature to start it spinning.

2. If the armature or coil still does not spin, recheck all of your connections to be sure the battery and wires are securely connected.

3. If the coil makes just a partial turn, the coil is unbalanced. To balance the coil, adjust the ends of the coil where it touches the paper clips. The coil must be balanced to operate correctly.

Electricity and Magnetism   Unit 9: Electric Motors

Name: _____ Date: _____

Investigation:
Now that your motor is running, it is time to do some investigation.

1. Predict what will happen if you add another battery in series to the current.
   _____
   _____
   _____
   _____

2. Add the second battery and check your prediction. Was your prediction correct? Explain.
   _____
   _____
   _____
   _____

3. Remove the second battery from the system. Predict what will happen if you add another magnet to the magnet below the armature.
   _____
   _____
   _____
   _____

4. Add the magnet; record your observations, and explain.
   _____
   _____
   _____
   _____

Conclusion:
After working with your electric motor, make a generalization as to its operation and how the motor works. Make a generalization as to how a motor works.

_____
_____
_____
_____
_____
_____
_____

Name: _____ Date: _____

Inquiry Investigation Rubric

Category	4	3	2	1
Participation	Used time well, cooperative, shared responsibilities, and focused on the task.	Participated, stayed focused on task most of the time.	Participated, but did not appear very interested. Focus was lost on several occasions.	Participation was minimal OR student was unable to focus on the task.
Components of Investigation	All required elements of the investigation were correctly completed and turned in on time.	All required elements were completed and turned in on time.	One required element was missing/or not completed correctly.	The work was turned in late and/or several required elements were missing and/or completed incorrectly.
Procedure	Steps listed in the procedure were accurately followed.	Steps listed in the procedure were followed.	Steps in the procedure were followed with some difficulty.	Unable to follow the steps in the procedure without assistance.
Mechanics	Flawless spelling, punctuation, and capitalization.	Few errors.	Careless or distracting errors.	Many errors.

Comments:

National Standards in Science, Math, and Technology

NSES Content Standards (NRC, 1996)
National Research Council (1996). *National Science Education Standards*. Washington, D.C.: National Academy Press.

UNIFYING CONCEPTS: K-12
Systems, Order, and Organization - The natural and designed world is complex. Scientists and students learn to define small portions for the convenience of investigation. The units of investigation can be referred to as systems. A system is an organized group of related objects or components that form a whole. Systems can consist of electrical circuits.

Systems, Order, and Organization
The goal of this standard is to ...
- Think and analyze in terms of systems.
- Assume that the behavior of the universe is not capricious. Nature is predictable.
- Understand the regularities in a system.
- Understand that prediction is the use of knowledge to identify and explain observations.
- Understand that the behavior of matter, objects, organisms, or events has order and can be described statistically.

Evidence, Models, and Explanation
The goal of this standard is to ...
- Recognize that evidence consists of observations and data on which to base scientific explanations.
- Recognize that models have explanatory power.
- Recognize that scientific explanations incorporate existing scientific knowledge (laws, principles, theories, paradigms, models) and new evidence from observations, experiments, or models.
- Recognize that scientific explanations should reflect a rich scientific knowledge base, evidence of logic, higher levels of analysis, greater tolerance of criticism and uncertainty, and a clear demonstration of the relationship between logic, evidence, and current knowledge.

Change, Constancy, and Measurement
The goal of this standard is to ...
- Recognize that some properties of objects are characterized by constancy, including the speed of light, the charge of an electron, and the total mass plus energy of the universe.
- Recognize that changes might occur in the properties of materials, position of objects, motion, and form and function of systems.
- Recognize that changes in systems can be quantified.
- Recognize that measurement systems may be used to clarify observations.

National Standards in Science, Math, and Technology (cont.)

Form and Function

The goal of this standard is to …
- Recognize that the form of an object is frequently related to its use, operation, or function.
- Recognize that function frequently relies on form.
- Recognize that form and function apply to different levels of organization.
- Enable students to explain function by referring to form, and explain form by referring to function.

NSES Content Standard A: Inquiry
- Abilities necessary to do scientific inquiry
  - Identify questions that can be answered through scientific investigations.
  - Design and conduct a scientific investigation.
  - Use appropriate tools and techniques to gather, analyze, and interpret data.
  - Develop descriptions, explanations, predictions, and models using evidence.
  - Think critically and logically to make relationships between evidence and explanations.
  - Recognize and analyze alternative explanations and predictions.
  - Communicate scientific procedures and explanations.
  - Use mathematics in all aspects of scientific inquiry.
- Understanding about inquiry
  - Different kinds of questions suggest different kinds of scientific investigations.
  - Current scientific knowledge and understanding guide scientific investigations.
  - Mathematics is important in all aspects of scientific inquiry.
  - Technology used to gather data enhances accuracy and allows scientists to analyze and quantify results of investigations.
  - Scientific explanations emphasize evidence, have logically consistent arguments, and use scientific principles, models, and theories.
  - Science advances through legitimate skepticism.
  - Scientific investigations sometimes result in new ideas and phenomena for study, generate new methods or procedures, or develop new technologies to improve data collection.

NSES Content Standard B: Physical Science (Transfer of Energy) 5–8
- Energy is a property of many substances and is associated with heat, light, electricity, mechanical motion, sound, nuclei, and the nature of a chemical; energy is transferred in many ways.
- Electrical circuits provide a means of transferring electrical energy when heat, light, sound, or chemical changes are produced.
- In most chemical and nuclear reactions, energy is transferred into or out of a system. Heat, light, mechanical motion, or electricity might all be involved in such transfers.

National Standards in Science, Math, and Technology (cont.)

NSES Content Standard D: Earth and Space Science 5–8
- Structure of the Earth System
  - The earth's magnetic field

NSES Content Standard E: Science and Technology 5–8
- Abilities of technological design
  - Identify appropriate problems for technological design.
  - Design a solution or product.
  - Implement the proposed design.
  - Evaluate completed technological designs or products.
  - Communicate the process of technological design.
- Understanding about science and technology
  - Scientific inquiry and technological design have similarities and differences.
  - Many people in different cultures have made, and continue to make, contributions.
  - Science and technology are reciprocal.
  - Perfectly designed solutions do not exist.
  - Technological designs have constraints.
  - Technological solutions have intended benefits and unintended consequences.

NSES Content Standard F: Science in Personal and Social Perspectives 5–8
- Science and Technology in Society
  - Science influences society through its knowledge and world view.
  - Societal challenges often inspire questions for scientific research.
  - Technology influences society through its products and processes.
  - Scientists and engineers work in many different settings.
  - Science cannot answer all questions, and technology cannot solve all human problems.

NSES Content Standard G: History and Nature of Science 5–8
- Science as a human endeavor
- Nature of science
  - Scientists formulate and test their explanations of nature using observation, experiments, and theoretical and mathematical models.
  - It is normal for scientists to differ with one another about interpretation of evidence and theory.
  - It is part of scientific inquiry for scientists to evaluate the results of other scientists' work.
- History of science
  - Many individuals have contributed to the traditions of science.
  - Science has been, and is, practiced by different individuals in different cultures.
  - Tracing the history of science can show how difficult it was for scientific innovators to break through the accepted ideas of their time to reach the conclusions we now accept.

National Standards in Science, Math, and Technology (cont.)

Standards for Technological Literacy (STL) ITEA, 2000
International Technology Education Association (2000). *Standards for Technological Literacy.* Reston, VA: International Technology Education Association.

The Nature of Technology
Students will develop an understanding of the:
1. Characteristics and scope of technology.
2. Core concepts of technology.
3. Relationships among technologies and the connections between technology and other fields of study.

Technology and Society
Students will develop an understanding of the:
4. Cultural, social, economic, and political effects of technology.
5. Effects of technology on the environment.
6. Role of society in the development and use of technology.
7. Influence of technology on history.

Design
Students will develop an understanding of the:
8. Attributes of design.
9. Engineering design.
10. Role of troubleshooting, research and development, invention and innovation, and experimentation in problem solving.

Abilities for a Technological World
Students will develop abilities to:
11. Apply the design process.
12. Use and maintain technological products and systems.
13. Assess the impact of products and systems.

The Designed World
Students will develop an understanding of and be able to select and use:
14. Medical technologies.
15. Agricultural and related biotechnologies.
16. Energy and power technologies.
17. Information and communication technologies.
18. Transportation technologies.
19. Manufacturing technologies.
20. Construction technologies.

National Standards in Science, Math, and Technology (cont.)

Principles and Standards for School Mathematics (NCTM), 2000

National Council for Teachers of Mathematics (2000). *Principles and Standards for School Mathematics.* Reston, VA: National Council for Teachers of Mathematics.

Number and Operations
Students will be enabled to:
- Understand numbers, ways of representing numbers, relationships among numbers, and number systems.
- Understand meanings of operations and how they relate to one another.
- Compute fluently and make reasonable estimates.

Algebra
Students will be enabled to:
- Understand patterns, relations, and functions.
- Represent and analyze mathematical situations and structures using algebraic symbols.
- Use mathematical models to represent and understand quantitative relationships.
- Analyze change in various contexts.

Geometry
Students will be enabled to:
- Analyze characteristics and properties of two- and three-dimensional geometric shapes and develop mathematical arguments about geometric relationships.
- Specify locations and describe spatial relationships using coordinate geometry and other representational systems.
- Apply transformations and use symmetry to analyze mathematical situations.
- Use visualization, spatial reasoning, and geometric modeling to solve problems.

Measurement
Students will be enabled to:
- Understand measurable attributes of objects and the units, systems, and processes of measurement.
- Apply appropriate techniques, tools, and formulas to determine measurements.

Data Analysis and Probability
Students will be enabled to:
- Formulate questions that can be addressed with data and collect, organize, and display relevant data to answer them.
- Select and use appropriate statistical methods to analyze data.
- Develop and evaluate inferences and predictions that are based on data.
- Understand and apply basic concepts of probability.

Science Process Skills

Introduction: Science is organized curiosity, and an important part of this organization includes the thinking skills or information-processing skills. We ask the question "why?" and then must plan a strategy for answering the question or questions. In the process of answering our questions, we make and carefully record observations, make predictions, identify and control variables, measure, make inferences, and communicate our findings. Additional skills may be called upon, depending on the nature of our questions. In this way, science is a verb, involving active manipulation of materials and careful thinking. Science is dependent on language, math, and reading skills, as well as the specialized thinking skills associated with identifying and solving problems.

BASIC PROCESS SKILLS:

Classifying: Grouping, ordering, arranging, or distributing objects, events, or information into categories based on properties or criteria, according to some method or system.

> Example – Using a magnet to sort a set of objects according to their magnetic properties. Testing a set of objects to determine their status as conductors or insulators.

Observing: Using the senses (or extensions of the senses) to gather information about an object or event.

> Example – Seeing and describing the setup of several circuits and noting the differences in a series circuit and a parallel circuit.

Measuring: Using both standard and nonstandard measures or estimates to describe the dimensions of an object or event. Making quantitative observations.

> Example – Using a magnet, several paper clips, and a ruler to determine the relative strength of a magnet.

Inferring: Making an interpretation or conclusion, based on reasoning, to explain an observation.

> Example – Interpreting where the hidden circuits might be in a circuit card or inferring as to where the circuits are in the classroom walls.

Communicating: Communicating ideas through speaking or writing. Students may share the results of investigations, collaborate on solving problems, and gather and interpret data, both orally and in writing. Using graphs, charts, and diagrams to describe data.

> Example – Describing an event or a set of observations; participating in brainstorming and hypothesizing before an investigation; formulating initial and follow-up questions in the study of a topic; summarizing data, interpreting findings, and offering conclusions; questioning or refuting previous findings; making decisions; using graphs to show the relative strength of several magnets.

Science Process Skills (cont.)

Predicting: Making a forecast of future events or conditions in the context of previous observations and experiences.

> Example – Predicting the relative strength of an electromagnet, based on the number of winds around an iron core (nail).

Manipulating Materials: Handling or treating materials and equipment skillfully and effectively.

> Example – Arranging equipment and materials needed to conduct an investigation, then using the materials to set up a series circuit and a parallel circuit.

Replicating: Performing acts that duplicate demonstrated symbols, patterns, or procedures.

> Example – Setting up a simple demonstration electric motor using the diagrams, directions, and materials provided.

Using Numbers: Applying mathematical rules or formulas to calculate quantities or determine relationships from basic measurements.

> Example – Computing the relative strength of an electromagnet relative to the voltage used in the system.

Developing Vocabulary: Specialized terminology and unique uses of common words in relation to a given topic need to be identified and given meaning.

> Example – Using context clues, working definitions, glossaries or dictionaries, word structure (roots, prefixes, suffixes), and synonyms and antonyms to clarify meaning.

Questioning: Questions serve to focus inquiry, determine prior knowledge, and establish purposes or expectations for an investigation. An active search for information is promoted when questions are used. Questioning may also be used in the context of assessing student learning.

> Example – Using what is already known about a topic or concept to formulate questions for further investigation, hypothesizing and predicting prior to gathering data, or formulating questions as new information is acquired.

Using Cues: Key words and symbols convey significant meaning in messages. Organizational patterns facilitate comprehension of major ideas. Graphic features clarify textual information.

> Example – Listing or underlining words and phrases that carry the most important details, or relating key words together to express a main idea or concept.

Science Process Skills (cont.)

INTEGRATED PROCESS SKILLS

Creating Models: Displaying information by means of graphic illustrations or other multisensory representations.

> Example – Drawing a graph or diagram, constructing a three-dimensional object, using a digital camera to record an event, constructing a chart or table, or producing a picture or diagram that illustrates information about the setup of a simple electrical device, such as an electric motor.

Formulating Hypotheses: Stating or constructing a statement that is testable about what is thought to be the expected outcome of an experiment (based on reasoning).

> Example – Making a statement to be used as the basis for an experiment: "The number of objects that can be picked up by an electromagnet is proportionate to the voltage of the system."

Generalizing: Drawing general conclusions from particulars.

> Example – Making a summary statement following analysis of experimental results: "The relative strength of an electromagnet is related to the number of wraps of wire around an iron core."

Identifying and Controlling Variables: Recognizing the characteristics of objects or factors in events that are constant or change under different conditions and that can affect an experimental outcome, keeping most variables constant while manipulating only one variable.

> Example – Listing or describing the factors that are thought to, or would, influence the flow of current in an electrical circuit.

Defining Operationally: Stating how to measure a variable in an experiment and defining a variable according to the actions or operations to be performed on or with it.

> Example – Defining such things as the strength of an electromagnet in the context of the materials and actions for a specific activity. Hence, the strength of an electromagnet may be measured by finding the number of washers that an electromagnetic system is able to support.

Science Process Skills (cont.)

Recording and Interpreting Data: Collecting bits of information about objects and events that illustrate a specific situation, organizing and analyzing data that has been obtained, and drawing conclusions from it by determining apparent patterns or relationships in the data.

> Example – Recording data (taking notes, making lists/outlines, recording numbers on charts/graphs, tape recordings, photographs, writing numbers of the results of observations/measurements) from the series of experiments to determine the strength of an electromagnet and forming a conclusion that relates trends in data to variables.

Making Decisions: Identifying alternatives and choosing a course of action from among alternatives after basing the judgment for the selection on justifiable reasons.

> Example – Identifying alternative ways to solve a problem through the utilization of a simple electrical circuit; analyzing the consequences of each alternative, such as cost or the effect on other people or the environment; using justifiable reasons as the basis for making choices; choosing freely from the alternatives.

Experimenting: Being able to conduct an experiment, including asking an appropriate question, stating a hypothesis, identifying and controlling variables, operationally defining those variables, designing a "fair" experiment, and interpreting the results of an experiment.

> Example – Utilizing the entire process of designing, building, and testing various electrical devices to solve a problem; arranging equipment and materials to conduct an investigation; manipulating the equipment and materials; and conducting the investigation.

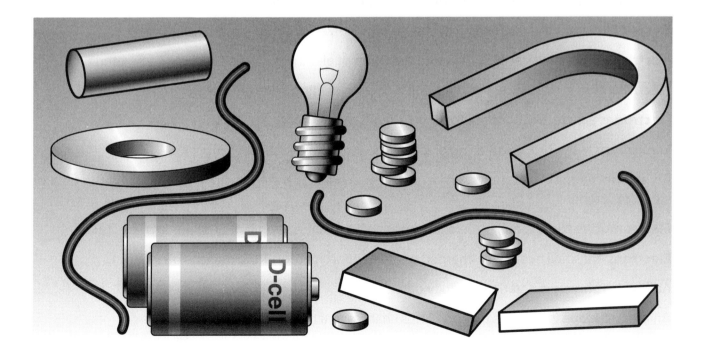

Definitions of Terms

Ampere is the measure for the unit of current.

An **armature** is the rotating part of an electric motor.

Atoms are tiny parts that make up all matter.

To **attract** is to cause to move together.

A **breaker** is a protective device in a circuit that will trip to prevent the build-up of heat.

The **Bohr model** is one way to explain the structure of an atom.

Circuit breakers are devices in a circuit that will either melt or trip to prevent the build-up of heat.

A **compass** is a freely suspended magnet, usually in the form of a magnetized needle that will align itself with the earth's magnetic poles.

Conduction is the transfer of charges (electrons) through direct contact with other objects.

A **conductor** is a material that allows an electric charge (electrons) to flow through it.

Current electricity is the movement of electrons, which creates the flow of electricity.

An **electrical charge** is a property of matter. Particles have either a positive (+) or negative (–) electrical charge.

An **electric circuit** is a closed-loop conducting path that consists of an energy source, an appliance or electric load, and wires that conduct the electric current from the energy source, through the appliance, and back to the energy source.

An **electric motor** is a device that converts electricity to mechanical work.

Electric current is the rate of flow of charge past a given point in an electric circuit.

Electricity is the physical attraction and repulsion of electrons within and between materials.

Electrification is the process of changing a body by adding or removing electrons.

Electrodes are electrical conductors.

Electrolysis is the process of using electricity to induce a chemical change.

An **electromagnet** is a temporary magnet in which the magnetic field is produced by the flow of electric current.

Electromotive force (emf) is the energy per unit of charge supplied by a source of electricity.

Electrons are small negatively charged particles in orbit around the nucleus of an atom.

An **electrophorus** is a static electricity detector.

An **electroscope** is a device that produces charges as a result of friction.

Definitions of Terms (cont.)

An **electrostatic charge** is a charge that is confined to an object; that is, the object has taken on or lost electrons, and therefore has a net negative or positive charge.

Electrostatic generators produce charges as a result of friction and the migration of electrons within and between materials.

Fuses are protective devices in a circuit that will melt to prevent the build-up of heat.

Friction produces charges through the rubbing of one object or surface against another.

A **galvanometer** is an instrument used for measuring small electric currents.

A **generator** is a device that produces electricity.

Humidity is water vapor in the air.

An **induced current** is when a magnet moving in the vicinity of a coil of wire generates an electric current.

Induction is the bringing about of a charge in other objects without contact, with no transfer of electrons.

An **insulator** is a material that restricts an electric charge from flowing through it. Some insulators may be considered nonconductors.

A **kilowatt** represents 1,000 watts.

A **kilowatt-hour (kWh)** is equal to the energy of 1,000 watts working for one hour.

Lightning is an atmospheric discharge of electricity, which typically occurs during thunderstorms.

A **lightning rod** is a metal lightning conductor used to protect wooden buildings.

A **line of flux (line of force)** is a line around a pole of a magnet representing the direction of magnetic field.

Lodestone (iron ore) is found naturally in the earth's surface and has magnetic qualities.

A **magnet** is a device that attracts certain metals, such as iron, nickel, and cobalt.

A **magnetic domain** is a region in which the magnetic fields of atoms are grouped together and aligned.

A **magnetic field** is a region in which a magnetic force can be detected.

A **magnetic force** is a force associated with the motion of electric charges and can be measured indirectly by finding out how many similar objects a magnet can pick up.

Magnetism is the alignment of electrons in the atom in the same direction, making regions called domains, creating a magnetic field.

Definitions of Terms (cont.)

Something with a **negative charge** is a material that has gained electrons.

The **nucleus** is the center of the atom.

Neutral is having no charge.

Neutrons are small particles that have no charge and are located in the nucleus of an atom.

Ohm is the measure for the unit of resistance in electric current.

Ohm's Law is the equation $E = IR$.

A **parallel circuit** is a circuit with two or more appliances that are connected to provide separate conducting paths for current for each appliance.

A **permanent magnet** is one made from a material that stays magnetized.

Poles are the two ends of a magnet.

Something with a **positive charge** is a material that has lost electrons.

Protons are small positively charged particles located in the nucleus of an atom.

To **repel** is to cause to move apart.

Resistance is the opposition to flow of electricity.

A **rotation** is a movement of an object in a circular motion.

A **series circuit** is a circuit with two or more appliances that are connected to provide a single conducting path for current.

Static electricity is defined as electricity at rest.

Switches are used to open and close circuits.

A **temporary magnet** is one that will lose its magnetism.

True north is the North Pole on a globe.

A **Van de Graaff generator** is a device that produces charges through friction.

Vehicles are different types of mechanical transportation.

A **volt** is the measure for the unit of potential difference.

Watts are units used to measure electricity.

A **Wimshurst machine** is a device that produces charges through friction.

Answer Keys

Historical Perspective
Quick Check (page 6)
Matching
1. c 2. d 3. e 4. a 5. b
Fill in the Blanks
6. static electricity   7. electrolysis
8. electric current
9. Michael Faraday, Joseph Henry
10. voltage, resistance, current
Multiple Choice
11. b 12. a 13. d

Electricity
Quick Check (page 12)
Matching
1. c 2. e 3. b 4. a 5. d
Fill in the Blanks
6. generator   7. protons, neutrons, electrons
8. electrical charge 9. conductors
10. kilowatt-hours
Multiple Choice
11. c 12. b 13. b
Knowledge Builder (page 13)
1. 5 hours   2. 0.5 or ½ hour
3. 2.5 or 2 ½ hours 4. 4 hours

Current Electricity
Quick Check (page 16)
Matching
1. e 2. a 3. b 4. d 5. c
Fill in the Blanks
6. Electrical circuits   7. Fuses, breakers
8. series, parallel   9. electrons
10. circuit
Multiple Choice
11. d 12. a
Knowledge Builder (page 17)
Activity #1
1. no 2. no 3. yes 4. yes 5. no 6. no
Observation:
1. silver tip of the base of the bulb
2. metal side of the bulb
Activity #2
1. the bulb lights   2. the bulb does not light
Inquiry Investigation (page 18–19)
Part I
Observation #1
1. Answers may vary but should be either series or parallel.
2. Parallel circuit: the brightness of the bulb will be the same as the original circuit. Series circuit: the bulb will be brighter.

Observation #2
1. Answers may vary but should be either series or parallel.
2. Parallel circuit: brightness of bulb will be the same. Series circuit: the bulb will be brighter.
Conclusion: The bulb will be brighter when two batteries are in series than when two batteries are in parallel.
Part II
Observation #1
1. Student's response will be either the bulb stays lit or the bulb goes out depending on the type of circuit constructed.
2. Answers will vary.
Observation #2: The bulbs are brighter in a parallel circuit than in a series circuit.
Conclusion:
1. When bulbs are in a series and one bulb burns out, then the other bulb stops functioning.
2. When bulbs are in parallel and one bulb burns out, then the other bulb will continue to be lit.

Static Electricity
Quick Check (page 23)
Matching
1. d 2. e 3. b 4. a 5. c
Fill in the Blanks
6. Protons, electrons   7. conduction
8. positive charge   9. attract
10. lightning rod
Multiple Choice
11. b 12. d 13. c 14. d
Knowledge Builder (page 24)
Activity #1—Observation: The water bends toward the balloon
Conclusion: The neutral water was attracted to the charged balloon and moved toward it.
Activity #2—Conclusion: In the bathroom, water in the air and on the walls helped move electrons away from the balloon quickly, so the balloon fell much quicker.
Inquiry Investigation (page 25–26)
Observation:
1. The plastic bag was rubbed against the plastic lid to create a charge.
2. Answers will vary but might include walking on a carpet, sliding across the seat of the car, combing my hair, removing a sweater, clothes rotating in the dyer, etc.
3. The Styrofoam or paper pieces are attracted to the lid because they have a different charge. (The charge on the Styrofoam or paper is induced by the lid. The lid has a strong positive charge because it has given up electrons to the plastic bag.)

4. The pieces at the top are negative, and the pieces at the bottom are positive.
5. The Styrofoam or paper pieces lose some of their electrons to the plastic lid and then are repelled by it as they become positively charged.

Conclusion: Answers may vary but should include the following information: Friction can cause an object to gain a charge. The plastic bag rubbing against the lid created a charge. The lid gave up an electron through direct contact with the plastic bag, which is conduction. The charge on the Styrofoam or paper is now induced by the lid. They are attracted to the lid. They lose electrons to the plastic lid and are repelled as they become positively charged.

Static Discharge
Quick Check (page 30)
Matching
1. b   2. a   3. d   4. e   5. c

Fill in the Blanks
6. Benjamin Franklin   7. Van de Graaff
8. Wimshurst   9. electrons
10. charge

Multiple Choice
11. c   12. a   13. d

Knowledge Builder (page 31)
Activity #1: Lightning happens when the negative charges, which are called electrons, in the bottom of the cloud, or in this experiment, on your finger, are attracted to the positive charges, which are called protons, in the ground, or in this experiment, on the aluminum pie pan. The resulting spark is like a mini lighting bolt.

Activity #2: The strips are charged the same way as each other and so they repel each other.

Inquiry Investigations (page 32–33)
Observation #1: sparks jump between pie pan and finger
Observation #2:
   A. no sparks   B. spark   C. no spark
   D. spark   E. spark
   Step 9: no spark   Step 10: no spark

Conclusion:
1. The handle serves as an insulator.
2. Electrons are transferred from the aluminum pie pan to the finger when the pan is sitting on the Styrofoam sheet and from the finger to the aluminum pan when the pan is raised.
3. The needle affects the ability of the aluminum pan to transfer electrons.

Magnets
Quick Check (page 37)
Matching
1. b   2. d   3. a   4. e   5. c

Fill in the Blanks
6. poles   7. north, south
8. north, south   9. magnetic force
10. magnetic field

Multiple Choice
11. a   12. a   13. c   14. b

Knowledge Builder (page 38)
Activity #1: Magnets are attracted to metal and not attracted to materials such as paper, wood, glass, and plastics.

Activity #2—Observation: The cereal will swing to touch the balloon. Then the cereal jumps away by itself. It moves away as the balloon approaches.

Conclusion: Rubbing the balloon moved electrons from your hair to the balloon. The balloon had a negative static charge. The neutral cereal was attracted to it. When they touched, electrons slowly moved from the balloon to the cereal. Now both objects had the same negative charge, and the cereal was repelled.

Investigation #1 (page 39)
Observation: Answers will vary.

Investigation #2 (page 40)
Step 3:

Observation:
2. Most of the filings go to the ends of the magnet with many lining up between the north and south pole; they line up in semi-circles, with their ends aligning with the north and south poles.
3. The floating needle floats toward the poles in paths similar to the lines formed by the magnetized iron filings.
4. The distance that the floating needle moves is similar to the distance that the iron filings are aligned. The needle is affected by the magnet from farther away than the iron filings; this is because there is less friction with the water.
5. Because there is a magnetic field around that magnet, and the needle and iron filings are magnetic materials. There is a magnetic field around the magnet, and the magnetic force acts in this area.
6. The strongest parts of the magnet are at each end where the north and south poles are located and the area in the center between the poles.
7. By sprinkling iron filings around other objects to determine if a magnetic field is present.

Conclusion: Magnetic field lines formed by iron filings represent one way to study magnetic fields around a magnet. The iron filings can help to determine the strength and direction of the magnetic field and the size of that field. The shape of the field appears to show concentric rings extending from pole to pole and outward from the poles.

Investigation #3 (page 41)
Conclusion: When magnets are combined, the new combination acts as one magnet with an observable increase in force.

Investigation #4 (page 42)
Conclusion: When magnets are combined, the new combination acts as one magnet with an observable increase in force.

Investigation #5 (page 43–44)
Part I: Answers will vary.
Part II: Answers will vary.
Conclusion: A magnet's force acts through space, and certain materials appear to be relatively transparent to a magnetic field. Magnetic materials such as iron or steel will interrupt or block the magnetic force. (The iron or steel is said to "cut" the magnetic field.) Iron or steel is a material that may be used to shield objects from a magnetic field.

A Compass and Earth's Magnetic Field
Quick Check (page 47)
Matching
1. c   2. d   3. a   4. b   5. e
Fill in the Blanks
6. magnet
7. magnetic field
8. geographic
9. north-seeking
10. deflected
Multiple Choice
11. c   12. b   13. b

Knowledge Builder (page 48)
Activity #1: Earth's magnetic field
Activity #2: North; yes
Conclusion:
1. The north pole of the magnet is attracted to the earth's magnetic north pole.
2. Plastic or glass materials do not interfere with the magnetic field.

Electromagnetism
Quick Check (page 51)
Matching
1. a   2. c   3. d   4. b   5. e
Fill in the Blanks
6. electromagnet, magnetism
7. Electricity, magnetism
8. iron
9. magnetic field
10. electromagnet
Multiple Choice
11. b   12. d   13. c

Knowledge Builder (page 52)
Conclusion: An electromagnet takes advantage of the relationship between magnetism and electricity. An advantage of an electromagnet over a permanent magnet is that it can be turned on and off, and the strength of the magnetism can be varied. One way to increase the strength of the electromagnet is by adding more wire wraps around the nail. The greater the number of wraps, the stronger the electromagnet.

Electric Motors
Quick Check (page 55)
Matching
1. b   2. d   3. e   4. a   5. c
Fill in the Blank
6. gears, pulleys
7. permanent
8. attraction, repulsion
9. electricity
10. efficient
Multiple Choice
11. b   12. c   13. d

Inquiry Investigation (page 56–60)
Investigation:
1. and 3. Possible predictions might include, but not be limited to, "there will be no effect on the spin of the motor"; "the motor may spin faster," etc.
2. and 4. Student responses will vary but should include an observation of how the motor changed and a reasonable explanation as to why the change occurred.
Conclusion: Student responses will vary. Response should indicate an understanding of the interaction of the magnetic fields created by the moving electrical current and the permanent magnet. Responses may also indicate a correlation between the speed of the motor and the amount of voltage supplied to the motor and a relationship between the number of permanent magnets and the motor's speed.

Bibliography

Children's Literature Resources:

Aczel, A.D. (2001). *The Riddle of the Compass: The Invention That Changed the World.* New York, N.Y.: Harcourt, Inc.

Berger, M. (1989). *Switch On, Switch Off.* New York, NY: HarperCollins Publishers.

Branley, F.M. (1996). *What Makes A Magnet?* New York, NY: HarperCollins Publishers.

Bryant-Mole, K. (1998). *Magnets.* Des Plaines, IL: Heinemann Interactive Library.

Chapman, P. (1976). *The Young Scientist's Book of Electricity.* Tulsa, OK: Educational Development Corp.

Cleary, B. (1983). *Dear Mr. Henshaw.* New York, NY: Bantam Doubleday Dell Publishing.

Cole, J. & Degen, B. (1997). *The Magic School Bus and the Electric Field Trip.* New York, NY: Scholastic Press.

Flaherty, M. (1999). *Science Factory: Electricity and Batteries.* Brookfield, CT: Copper Beech Books.

Gordon, M. (1996). *Electricity and Magnetism.* New York, NY: Thomson Learning.

Gunderson, P.E., (1999). *The Handy Physics Answer Book.* Farmington Hills, MI: Visible Ink Print.

Lafferty, P. (1989). *Magnets to Generators.* New York, NY: Gloucester Press.

Pinchuk, A. (2001). *Popular Mechanics for Kids: Make Amazing Toy and Game Gadgets.* New York, NY: HarperCollins Publishers.

Pinchuk, A. (2001). *Popular Mechanics for Kids: Make Cool Gadgets for Your Room.* New York, NY: HarperCollins Publishers.

Riley, P. (1999). *Magnetism.* Danbury, CT: Franklin Watts.

Tocci, S. (2001). *Experiments with Electricity.* New York, NY: Children's Press.

Tocci, S. (2001). *Experiments with Magnets.* New York, NY: Children's Press.

Whalley, M. (1994). *Experiments with Magnets and Generators.* Minneapolis, MN: Lerner Publications Co.

Bibliography (cont.)

Curriculum and Technology Resources:

Software:

Electricity CD-ROM. (2000). Bethesda, MD: Discovery Channel Science.

Magnetism CD-ROM. (2000). Bethesda, MD: Discovery Channel Science.

Macauley, D. (1998). *The New Way Things Work* CD-ROM. New York, NY: Dorling Kindersley.

Science Court: Electric Current. (1998). Watertown, MA: Tom Snyder Productions.

Science Court Explorations: Magnets. (1999). Watertown, MA: Tom Snyder Productions.

Super Solvers: Gizmos & Gadgets. (1994). Fremont, CA: The Learning Company.

ZAP: The Science of Light, Sound, and Electricity. (1998). Redmond, WA: Edmark Corporation.

Web Resources:

http://jersey.uoregon.edu/vlab/Voltage/

http://www.mos.org/sln/toe/staticmenu.html

http://sln.fi.edu/franklin/activity.html

http://galileo.phys.virginia.edu/classes/620/electricity_Home.html

http://www.exploratorium.edu/snacks/eddy_currents/index.html

http://www.eskimo.com/~billb/emotor/statelec.html

Bibliography (cont.)

Curriculum Resources:

Aczel, A.D. (2001). *The Riddle of the Compass: The Invention That Changed the World.* New York, N.Y.: Harcourt, Inc.

Allen, D.S. and Ordway, R.J. (1960). *Physical Science.* Princeton, N.J.: D. Van Nostrand Company, Inc.

Allen, M., Bredt, D., Calderwood, J., Chambers, P., Deal, D., Hoover, E., Kahn, G. P., Kirkhart, J., Larimer, H., Mercier, S., Schmeling, S., Sipkovich, V., & Walsh, M. (1991). *Electrical Connections, AIMS Activities for Grades 4-9.* Fresno, CA: AIMS Education Foundation.

Marson, R. (1983). *Electricity.* Canby, OR: TOPS Learning Systems.

Marson, R. (1983). *Magnetism.* Canby, OR: TOPS Learning Systems.

Metcalfe, H.C., Williams, J.E. and Dull, C.E. (1960). *Modern Physics.* New York, N.Y.: Henry Holt and Co, Inc.

National Academy of Sciences. (2002). *Electric Circuits.* Burlington, NC: Carolina Biological Supply Company.

National Academy of Sciences. (2002). *Magnets and Motors.* Burlington, NC: Carolina Biological Supply Company.

Schafer, L.E. (2001). *Charging Ahead: An Introduction to Electromagnetism.* Washington, D.C.: National Science Teachers Association.

Schafer, L.E. (1992). *Taking Charge: An Introduction to Electricity.* Washington, D.C.: National Science Teachers Association.

Operation Physics Electricity. (circa 1989). *Electricity.* American Institute of Physics.

Operation Physics Electricity. (circa 1989). *Magnetism.* American Institute of Physics.